I0493932

NISTIR 7546

Airtightness Evaluation of Shelter-in-Place Spaces for Protection Against Airborne Chembio Releases

Andrew Persily
Heather Davis
Steven J. Emmerich
W. Stuart Dols
Building and Fire Research Laboratory

Prepared for:
U.S. Environmental Protection Agency
Research Triangle Park, NC
Funded under IAG DW-13-92178301-0 by the U.S. EPA National Homeland Security Research Center (NHSRC), Decontamination and Consequence Management Division (DCMD)

October 2009

U.S. Department of Commerce

Gary Locke, *Secretary*

National Institute of Standards and Technology

Patrick D. Gallagher, *Deputy Director*

ABSTRACT

Due to concerns about potential airborne chemical and biological releases in or near buildings, building owners and managers and other decision makers are faced with a number of options for increasing their building's level of protection against such events. Among the various technologies and approaches being proposed and implemented is shelter-in-place (SIP). SIP strategies involve having the building occupants stay in the building, generally in a space designated for such sheltering, until the event is over and the outdoor contaminant levels have decreased. While much guidance is available on the implementation of SIP in buildings, important technical issues remain about the degree of protection provided by a particular space and the factors in determining the level of protection. In particular, many recommendations suggest tightening the walls of SIP spaces, but there has been insufficient analysis of the relationship between shelter tightness and the protection provided by the SIP space.

In order to address some of these questions, the National Institute of Standards and Technology (NIST) has undertaken a project for the U.S. Environmental Protection Agency to develop and demonstrate evaluation methods to relate shelter airtightness to the performance of shelter-in-place approaches for airborne chemical, biological and radiological (CBR) protection of building occupants. The focus of this effort is on short term sheltering, on the order of hours, rather than longer term sheltering which generally employ filtration and air cleaning equipment to supply clean air to the occupants of the space. This project has consisted of the following tasks: a literature review of SIP strategies and performance issues; development of a study plan for testing SIP airtightness evaluation methods; implementation of the study plan through a combination of experiments and simulations; and, finally, development of recommendations on SIP evaluation and possible performance criteria for candidate SIP spaces.

Keywords: building protection, CBR, chembio, shelter-in-place, SIP

Table of Contents

LIST OF ACRONYMS

AC: air conditioning

ACGIH: American Conference of Governmental Industrial Hygienists

ASHRAE: American Society of Heating, Refrigerating and Air-Conditioning Engineers

BFRL: Building and Fire Research Laboratory (part of NIST)

CBR: chemical, biological and radiological

CP: collective protection

CONTAM: NIST-developed multizone airflow and contaminant dispersal simulation program

CSEPP: Chemical Stockpile Emergency Preparedness Program (FEMA program)

DRF: dose reduction factor

ELA effective leakage area

EPA: Environmental Protection Agency

FEMA: Federal Emergency Management Agency

HVAC: heating, ventilating and air-conditioning

IAQ: indoor air quality

MERV: minimum efficiency reporting value, metric of particle filtration efficiency based on ASHRAE Standard 52.2

NIOSH: National Institute of Occupational Safety and Health

NIST: National Institute of Standards and Technology

ORNL: Oak Ridge National Laboratory

PF: protection factor

SIP: shelter-in-place

TLV: threshold limit value

1. INTRODUCTION

Due to concerns about potential airborne chemical and biological releases in or near buildings, building owners and managers and other decision makers are faced with a number of options for increasing their building's level of protection against such events [1]. A wide range of technologies and approaches are being proposed with varying levels of efficacy and cost, as well as varying degrees of applicability to any particular building. In particular, shelter-in-place (SIP) has been proposed as a strategy to protect building occupants from chembio releases, particularly outdoor releases. SIP strategies involve having the building occupants stay in the building, generally in a space designated for such sheltering, until the event is over and the outdoor contaminant levels have decreased and it is safe to leave the building. SIP is often considered as an alternative to building evacuation under conditions where the outdoor exposure is likely to be higher than the exposure in the shelter. While much guidance is available on the implementation of SIP in buildings [2], important technical issues remain about the degree of protection provided by a particular space and the factors that determine the level of protection. Additional questions exist regarding the appropriate duration of occupancy based on concerns regarding oxygen depletion, carbon dioxide buildup and exposure to the chembio agent over time, as well as the role of filtration and air cleaning in providing additional protection by pressurizing the SIP space with clean air. Also, while many recommendations suggest tightening the partitions to adjacent spaces, there has been insufficient analysis of the relationship between the shelter tightness and the protection provided by the SIP space nor any recommended quantitative airtightness criteria.

In order to address some of these questions, the National Institute of Standards and Technology (NIST) has undertaken a project for the U.S. Environmental Protection Agency to develop and demonstrate evaluation methods to relate shelter airtightness to the performance of shelter-in-place approaches for CBR protection of building occupants. The focus of this effort is on short term sheltering, on the order of hours, rather than longer term sheltering which generally employs filtration and air cleaning equipment to supply clean air to the occupants of the space. However, the results of this effort still have application to longer term sheltering as the airtightness of these shelters determines the amount of clean air supply required to maintain the shelter at positive pressure.

This project has consisted of the following tasks: a literature review of SIP strategies and performance issues; development of a study plan for testing SIP airtightness evaluation methods; implementation of the study plan through a combination of experiments and simulations; and, finally, development of recommendations on SIP evaluation and possible performance criteria for candidate SIP spaces.

This report is organized by these tasks with the first section presenting the results of the literature review. The next section describes the experimental and other analysis approaches used in this study, followed by a section with the results of those efforts. A series of recommendations for SIP space evaluation is presented in the final section of the report.

2. LITERATURE REVIEW

In order to support the technical work involved in this project, a review of the existing literature on shelter-in-place was conducted, including several guidance documents intended for practitioners, to identify what measures have been proposed to evaluate candidate SIP spaces for the degree of protection they offer against an outdoor release. While the review was focused primarily on finding quantitative evaluation methods to judge candidate spaces, it was also an opportunity to collect qualitative considerations proposed for SIP spaces.

1

This review covered a range of documents, which are divided into three categories: Research Reports and Papers; Chemical Stockpile Emergency Preparedness Program (CSEPP), and Other Federal Emergency Management Agency (FEMA) documents; and other guidance documents. The first category includes the results of several research studies and analyses into the effectiveness of shelter-in-place protection strategies, including some experimental studies. The second category contains a number of reports produced under the FEMA CSEPP program, which exists to produce information, guidance and training material "for formulating and coordinating emergency plans and the associated emergency response systems for chemical events that may occur at the chemical agent stockpile storage locations in the continental United States." The third category of documents reviewed consists of SIP guidance material produced by a number of organizations, both for public education and for building and system design.

2.1 Research Reports

SIP research provides an indication of the protection offered by sheltering within a building, in some cases addressing the additional protection provided by building tightening or by filtration and air cleaning. This work includes field tests [3-6], theoretical analyses [7-9] and simulation studies [10, 11]. The field tests tend to focus on the sheltering provided by a building, and not on the additional sheltering provided by an SIP space within a building. Jetter and Whitfield [12] examined the protection provided by an interior bathroom of a residence, in which tracer gas tests were used to estimate airflow rates between the shelter, the rest of the house and the outdoors. In a more recent study, these airflows were then used to estimate protection factors (ratio of outdoor to shelter exposure) as a function of time. Jetter and Profitt conducted similar tests in commercial buildings, which are discussed in more detail later in this report [13].

The theoretical work provides insight into the parameters that determine the protection provided by the building as a whole (i.e., airtightness, weather conditions, and particle deposition rates), but does not focus on SIP spaces. By considering the mass balance for a single zone, an equation relating the dose reduction factor (DRF) relative to outdoors was derived by Engelmann [7, 8] for a step change in the outdoor concentration from zero to a nonzero value, as follows:

$$\text{DRF} = \text{Indoor exposure/Outdoor exposure} = 1 - (1/Rt) + (e^{-Rt})/Rt \qquad (1)$$

where t is the time elapsed since the outdoor concentration increase and R is the building air change rate in units of air changes per hour or h^{-1}. Therefore, the lower the value of DRF, the lower the occupant exposure relative to outdoors.

Figure 1 is a plot of DRF calculated with equation (1) for a range of values of the building air change rate. Note that as t increases, the indoor exposure approaches the outdoor exposure and the DRF approaches 1. Also, lower air change rates correspond to lower values of DRF, i.e., lower indoor exposure, but the values of DRF still approach 1 after a sufficient amount of time elapses. While the plot shows that DRF approaches 1 regardless of the air change rate, the period of interest for most SIP applications discussed in this report is only on the order of a few hours, in which case the different air change rates correspond to very different DRF values. Similar analysis yields equations accounting for particle deposition and a subsequent step decrease in the outdoor concentration.

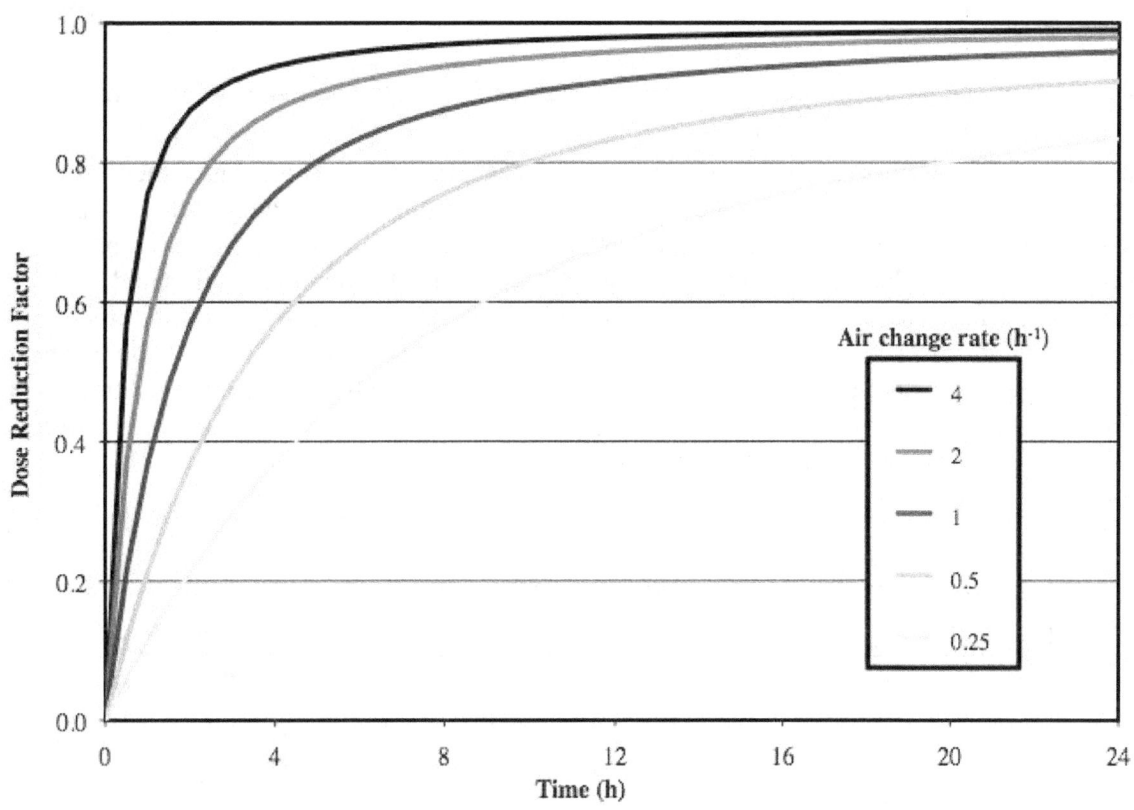

Figure 1 Dose reduction factor (DRF) as a function of time and air change rate

2.2 CSEPP and Other FEMA Documents

The CSEPP program, sponsored by FEMA, has produced a significant amount of material relevant to SIP strategies, much of it produced at Oak Ridge National Laboratory (ORNL). The resulting publications include guidance documents, experimental studies and other materials. One such publication, Blewett et al. [14], is fairly comprehensive, including a literature review, a discussion of different sheltering approaches, qualitative criteria for room selection, and the results of sheltering tests in twelve buildings. The four approaches described in this reference, as well as many other publications, include the following:

> *Normal* sheltering: closing all windows and doors, and turning off all mechanical equipment, such as HVAC systems
> *Expedient* sheltering: applying temporary air sealing measures to a shelter space, such as taping over vents or placing plastic sheeting over windows
> *Enhanced* sheltering: applying permanent air sealing measures to a shelter space
> *Pressurized* sheltering: providing filtered/cleaned air to the shelter to achieve an elevated air pressure relative to outside the shelter, thereby greatly limiting the entry of air and contaminant

This report and others speak in terms of the Protection Factor (PF), which is the outdoor dosage divided by the indoor dosage at some point in time and therefore equal to the inverse of the DRF defined above. Figure 2 is a plot of the Protection Factor as a function of time, analogous to the DRF plot in Figure 1. (Note that PF is higher for lower air change rates and shortly after the release begins, tending towards a value of 1.0 as time continues. Again the plot extends for many hours to show that all the curves tend towards a value of 1, but the period of interest for most applications is on the order of only a few hours.)

The experimental portion of the Blewett et al. study [14] was focused on measurement of the airflow rate between expediently sealed safe rooms and the outdoors, as well as the air change rate of the whole house. The results of these measurements were used to estimate a range of protection factors under normal and expedient sheltering associated with a 10 min and a 1 h outdoor exposure. The 10 min values of PF range from about 20 to 60 for the whole building, while the values of PF for expedient sheltering in a bathroom were almost twice as high. The 1 h values ranged from about 4 to 12 for the whole building and about 5 to 15 for the bathrooms. The report also summarizes some SIP guidance available at the time of the report and provides some qualitative criteria for selecting a SIP space in a building.

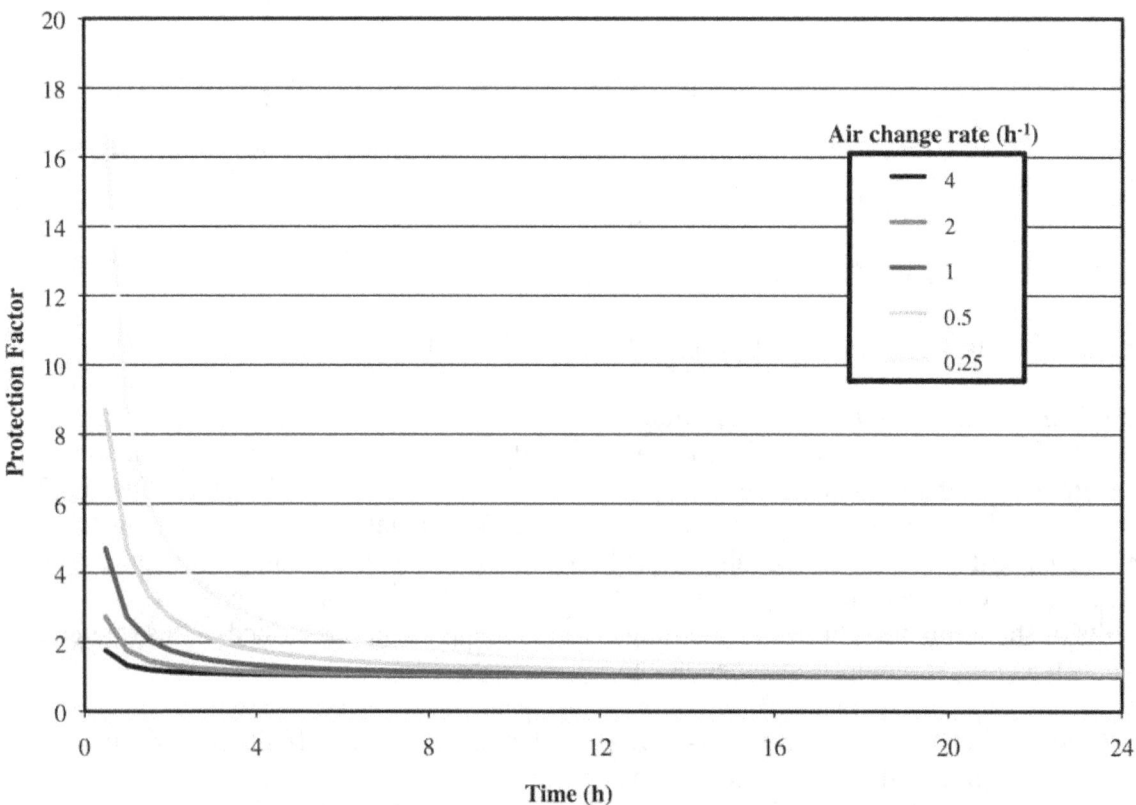

Figure 2 Protection factor as a function of time and air change rate (DRF_PF-graph.xls)

The CSEPP has also produced guidance documents on specific sheltering approaches [15, 16]. Other reports out of the CSEPP program are more policy related, discussing issues such as planning for SIP, deciding between evacuation and sheltering under a given scenario, managing building occupants [17-22] and assessing the housing stock near chemical storage sites [23]. This latter document looks primarily at the age of houses near these sites and discusses airtightness in general, but does relate building age to the airtightness of these houses. Other FEMA documents provide useful information, such as recommendations on floor area per person in tornado shelters [24].

More recently, FEMA issued design guidance for shelters and safe rooms [2]. While much of this report is devoted to issues of protecting building occupants from the effects of blast, it also contains important information on protection from chembio and radiological (CBR) airborne contaminants. FEMA distinguishes between three levels of CBR protection in shelter, the first being pressurization of the space combined with particle filtration and/or gaseous air cleaning. The second class includes filtration and/or air cleaning but with little or no pressurization, and class 3 is passive, meaning no air treatment or efforts to pressurize the shelter space. The document speaks to airtightening of shelters, either temporarily when an event occurs or permanently as part of a preparedness effort. The guide notes that no airtightness criteria exist for shelters but does note some key air leakage sites for sealing efforts and describes the use of fan pressurization or blower door testing as a means of quantifying shelter airtightness.

2.3 Other Guidance Documents

In addition to the CSEPP program, other organizations have issued guidance on the use of SIP as an exposure reduction strategy, ranging from general guidance [25-30] to more detailed design specifications [31, 32]. The latter two documents from the U.S. Army Corps of Engineers contain detailed design requirements for collective protection (CP), identifying four classes of facilities based on their potential to support integration of CP systems. These classes range from those with HVAC systems that are capable of integrating an overpressure system to those that cannot be pressurized without extensive sealing. These Army Corps documents also speak to floor area per person, air locks, envelope leakage testing, and filtration systems in a fairly detailed fashion. While these documents do not contain airtightness specifications, they do call for a minimum overpressure of 75 Pa for the Class I collective protection and 5 Pa for Class II. Some SIP guidance is also contained within more general discussions of building protection strategies [33].

2.4 Summary of Review

While the existing literature and guidance on SIP is well established and useful, it does not provide quantitative methods for evaluating the degree of protection that candidate SIP spaces might offer in the event of an outdoor exposure. Useful qualitative guidance on selecting such spaces exists, but there is little quantitative guidance other than recommendations on the amount of floor area per person.

The quantitative material that does exist employs the Dose Reduction Factor (DRF) and its inverse, the Protection Factor (PF), as measures of the degree of protection offered by a shelter. These parameters are defined in terms of the outdoor exposure relative to the exposure in the building or SIP zone. One could make the case that the SIP zone exposure should instead be compared to that in the rest of the building, as opposed to the outdoors, given that the occupants might stay in the building if no SIP space were available. Experimental studies have been conducted to evaluate SIP protection offered by whole buildings or specific building spaces, but these have tended to be tracer gas measurements of air change rates of the shelters, which do not relate directly to exposure reduction. Also, tracer gas testing methods are too involved and costly for most building owners and managers.

As noted, there are many useful qualitative recommendations for selecting a shelter space, addressing, for example size, location, and accessibility, with the recent FEMA guide [2] providing a very thorough description. Blewett et al. [14] includes the following attributes of SIP spaces in buildings: above ground, interior room with few or no windows, no plumbing fixtures if possible, no window AC units, at least 0.9 m² (10 ft²) per person, and not rooms with an exhaust fan linked to a light switch. Price et al. [33] note that the goal for an SIP space is to create a zone where infiltration is very low, which usually means being located in the interior portion of a building, i.e. no windows, and having doors that are fairly effective at preventing airflow from hallways. They also note that bathrooms are usually a bad choice because exhaust ducts can draw air into the room when there are stack driven airflows (i.e., arising from indoor-outdoor temperature differences) in a building, even when the fans are off. These various qualitative criteria for selecting SIP spaces are summarized in Section 5 of this report.

While most of the SIP guidance material recommends minimal air change with the outdoors and perhaps an effort to increase the airtightness of the shelter space, none of these documents makes specific recommendations on airtightness levels and very few address the measurement of shelter air leakage.

3. EVALUATION APPROACH AND STUDY PLAN

While the existing literature acknowledges the importance of shelter airtightness in protecting occupants against outdoor releases, it does not describe SIP space airtightness evaluation methods in any detail, nor does it present airtightness criteria. In order to address this need, the current project has pursued the evaluation of potential SIP spaces by measuring shelter airtightness with the fan pressurization method. The concept behind such an evaluation is that the shelter airtightness is related to the airflows between the SIP space, the rest of the building and the outdoors, and that these airflows are used to determine the contaminant levels in the shelter, the exposure of the shelter occupants to an outdoor release and ultimately the protection factor. The airflows, contaminant levels and occupant exposures for a given event will depend on the details of the release (e.g., the time profile and location), building and system operating conditions (outdoor air intake and other system airflows), outdoor weather conditions, and the interzone airflow dynamics within the building. However, as will be seen, the measured shelter airtightness can still be used to provide a reasonable indicator of the level of protection offered by a given shelter.

In order to investigate the pressurization-testing approach to SIP space evaluation, the following efforts were pursued under this project:

- SIP zone pressurization tests in several spaces to determine a range of airtightness values;
- Computer simulations to relate SIP zone airtightness to Protection Factor;
- Computer simulations to assess the ability to reliably measure SIP zone airtightness with fan pressurization; and
- Validation of predictions from shelter airtightness values.

This section describes the various steps pursued in investigating the pressurization test approach to SIP space evaluation, beginning with a discussion of the relevant theory and calculation methods.

3.1 Theory and Calculation Approaches

In order to relate SIP and building airflows to exposure, a two-zone mass balance theory is presented. This theory is employed by Jetter and Whitfield [12] and others, and is based on the key assumption that the shelter exchanges air only with the rest of the building volume and has no direct airflow connections with the outdoors. Similar theory can be developed for the more general case in which the shelter does exchange air directly with the outdoors. Another key assumption is that both the shelter and the remainder of the building can each be represented by a single value of the contaminant concentration. This latter assumption will often be reasonable for the shelter, particularly for a small space, but may be less justifiable for the rest of the building, particularly a large and complex building.

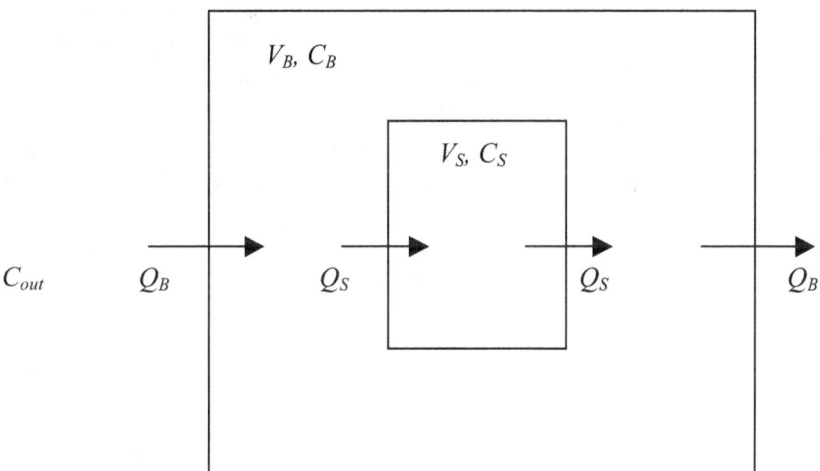

Figure 3 Schematic of two-zone SIP model

Figure 3 is a schematic of such a two-zone model, which is described by two mass balance equations of contaminant, one for each zone:

$$V_S \frac{dC_S}{dt} = Q_S C_B - Q_S C_S \tag{3}$$

$$V_B \frac{dC_B}{dt} = Q_B C_{OUT} + Q_S C_S - Q_S C_B - Q_B C_B \tag{4}$$

where,

V_S = volume of the shelter
V_B = volume of the rest of building
C_S = contaminant concentration in the shelter
C_B = concentration in the rest of the building
C_{OUT} = concentration outside the building
Q_S = airflow between the shelter and the building
Q_B = airflow between the building and outside, and
t = time.

Solving for C_S and C_B allows one to calculate exposure in the building and the shelter for specific volumes, airflows and outdoor release profiles. Alternatively, one can use this two-zone mass balance theory to analyze tracer gas test data to determine the airflows of interest, as was done by Jetter and Whitfield [12] in a low-rise residential building for which an interior bathroom served as the shelter.

Figure 4 is a sample plot of shelter and building concentrations for a simple single zone building with an SIP space subject to a 1 min elevation of the outdoor concentration to a dimensionless value of 1.0. The values in this plot were generated using the CONTAM model [34] for a simple two-zone model with weather-driven infiltration. The building concentration responds relatively quickly, while the shelter response lags that of the building. After the building concentration peaks, it then decreases while the shelter concentration continues to increase. For this case, the shelter concentration peaks after about 30 h at a value about one tenth of the building peak. From this point on, the shelter concentration is higher than the building concentration. The timing of the two peaks and their relative magnitudes are a function of the volumes and airflow rates in the two-zone model, as well as the time profile of the outdoor concentration. Nevertheless, the shelter peak will always be lower than the building peak but of longer duration. The fact that the shelter concentration eventually exceeds the building concentration demonstrates the need to leave the shelter after the outdoor threat has passed in order to avoid this longer term exposure in the shelter.

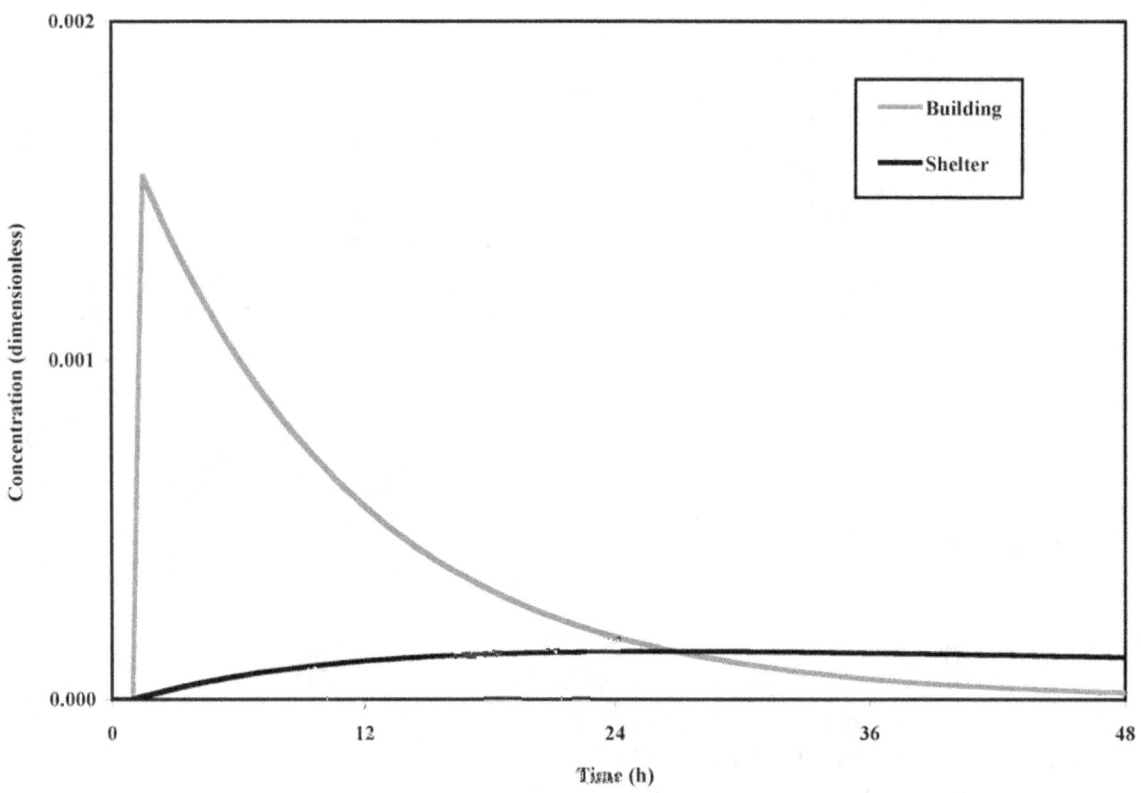

Figure 4 Contaminant concentration values for two-zone example

Figure 5 shows the Protection Factor for the shelter as a function of time for the same case as shown in Figure 4. Two values of Protection Factor are presented, one referenced to the outdoor exposure and the other referenced to the building exposure. Note that in both cases the shelter provides a high level of protection relative to the rest of the building early during the event, but the protection decreases as time progresses. The protection factors referenced to the building are lower than those referenced to outdoors due to the contaminant that enters the building. Eventually, both protection factors will decrease to a value of one as the indoor exposure eventually attains a value equal to the outdoor exposure.

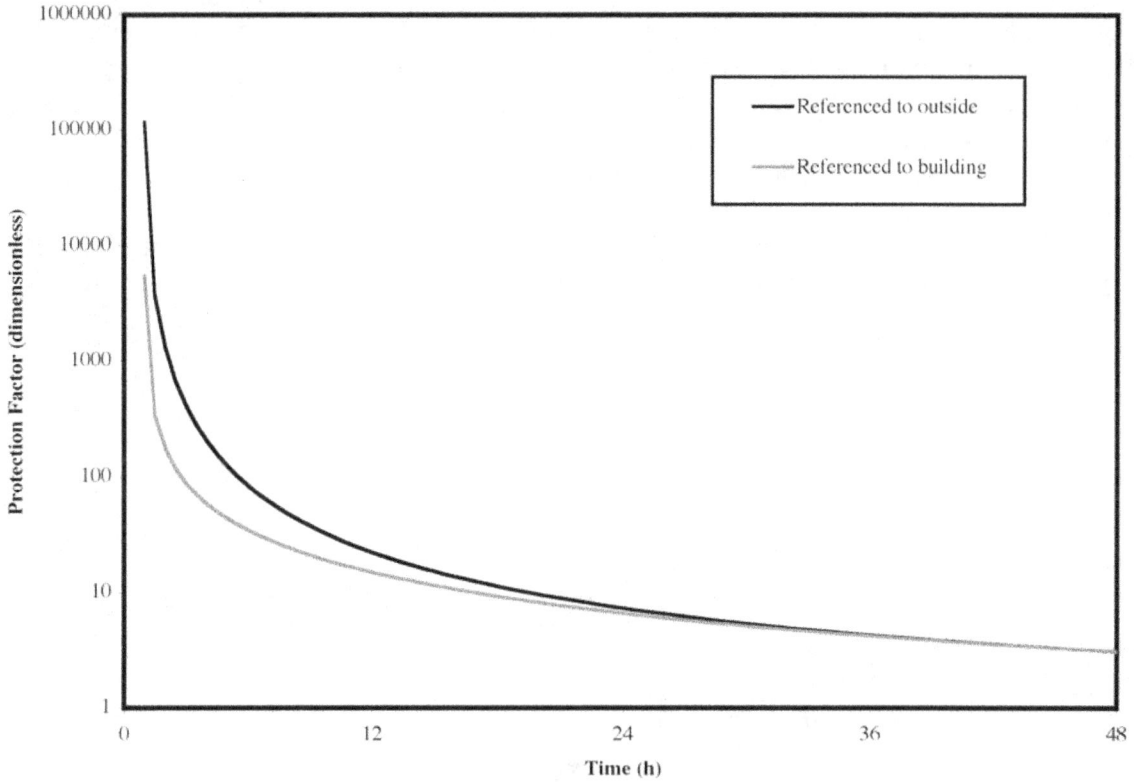

Figure 5 Protection factor values for two-zone example

3.2 SIP Zone Airtightness Tests

While some data exist for interior partitions in buildings [35], they are limited and not necessarily appropriate for spaces that have been sealed to provide sheltering. Therefore, this portion of the study involved using fan pressurization testing to determine a range of shelter airtightness values with and without expedient sealing efforts. The first step in this effort is to establish a test protocol and apply it to several potential shelter spaces. The airtightness tests are based on established fan pressurization methods, sometimes referred to as blower door testing. Fan pressurization testing employs a fan to mechanically pressurize or depressurize a room or building, while simultaneously measuring the airflow rate through the fan required to maintain the induced pressure difference. The tests in this study employed the procedures in ASTM Standards E779 and E1827 [36, 37], with some modifications since these standards are intended for testing whole buildings as opposed to rooms within a building.

Figure 6 shows a schematic of the test configuration for pressure testing an SIP space. The fan pressurization device is installed in a door to the SIP space and used to induce airflow into (out of) the space from (to) an adjacent interior space. Often a hallway serves as this adjacent space. The fan airflow raises (lowers) the pressure in the space relative to the adjoining space(s), with the airflow into (out of) the space leaving (entering) the SIP space through leakage paths connecting to adjoining spaces. In fan pressurization testing, the test is conducted at pressure differences from approximately 10 Pa to 70 Pa. The air leakage characteristics of the space or building being tested are calculated from the measured airflow rates and pressure differentials. A blower door device, shown below in Figure 7, was used for these tests. The airflow rates were determined from the pressure difference across an orifice plate built into the blower door using the manufacturer's calibration, which was confirmed in a blower door calibration chamber at NIST conducted in accordance with ASTM E1258 [38]. The blower door airflow rate has an associated uncertainty, based on these calibrations, on the order of 0.01 m^3/s. The pressure difference between the SIP zone and the adjacent hallway were recorded during the tests with a digital manometer that has a stated uncertainty of +/- 1 Pa.

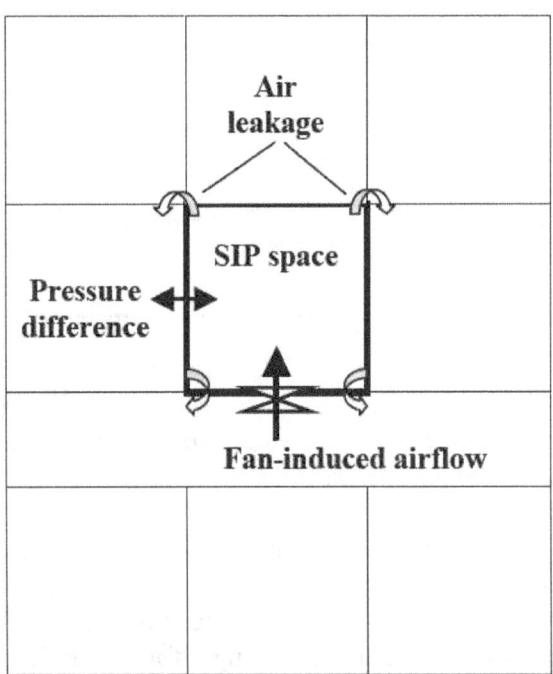

Figure 6 SIP pressurization test schematic

Figure 7 Blower door

These fan pressurization tests yield a series of pressure differences and airflow rates, an example of which is plotted in Figure 8. These data are then fit to a curve of the following form:

$$Q = C \, \Delta p^n \qquad (5)$$

where

Q = airflow rate, m^3/s
C = flow coefficient determined by curve fit, $m^3/s \bullet Pa^n$
Δp = pressure difference across the SIP zone, Pa
n = pressure exponent determined from curve fit, dimensionless

Once the pressurization test data are fit to a curve of the form shown in Equation (5), the curve is used to estimate the effective leakage area (ELA) of the space as a measurement of airtightness. The ELA is defined as the size of an orifice of discharge coefficient 1.0 that yields the same airflow rate as that predicted by the curve fit at some reference pressure, in this case 4 Pa [39]. The ELA values are then normalized by the surface area of the SIP space, including the walls, floor and ceiling to yield a value in units of cm^2/m^2.

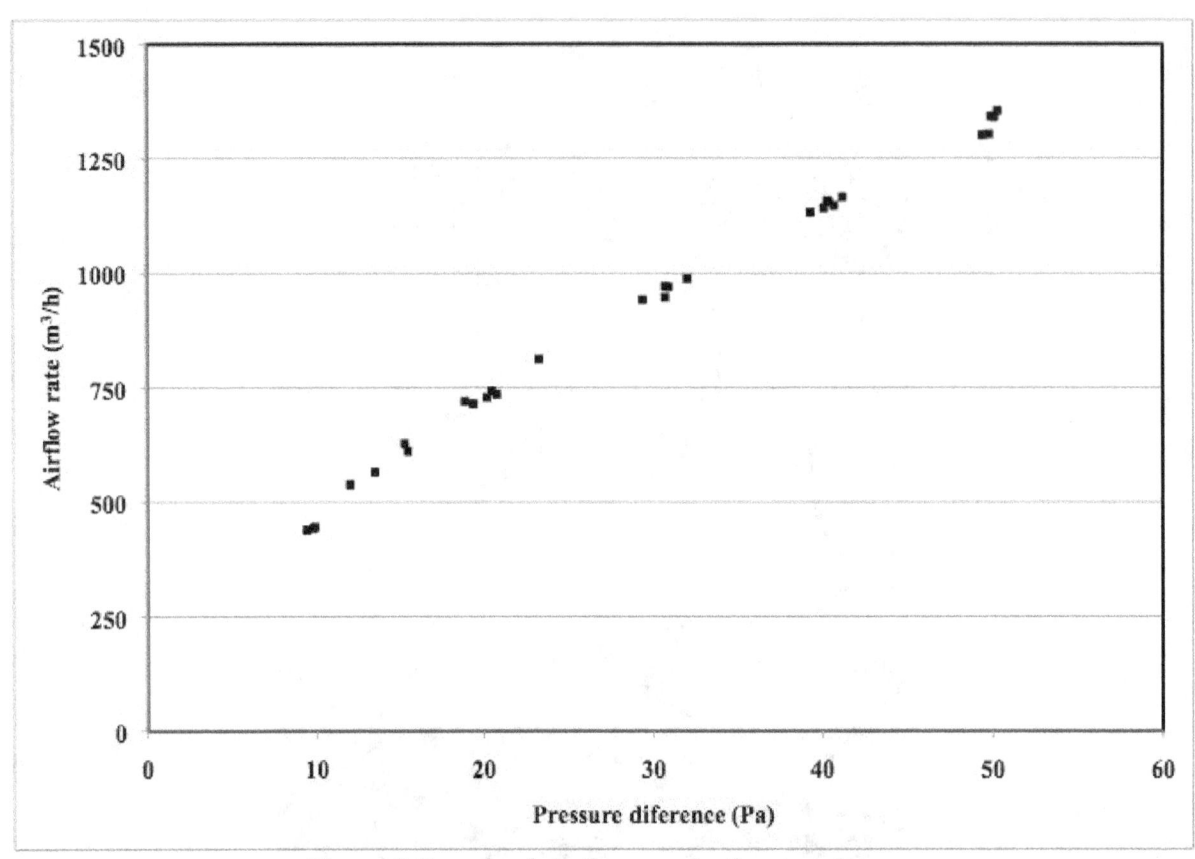

Figure 8 Sample plot of pressurization test data

SIP airtightness tests were performed in eight spaces in three buildings, as described in Table 1. All three buildings were located in suburban office park settings, with the first (226) building a roughly 40 year old, 3-story building on the NIST campus. The second building (NC-Off) is a 21 year old, single-story building located in the Research Triangle Park, NC, area, and the third (RTP) is a fairly new six-story building on the EPA RTP campus. All of the rooms tested had no exterior windows or walls, with one exception as noted below.

Room	Building	Nominal room dimensions* (m)
B221	226	8.1 x 6.5 x 3.4
B113	226	8.1 x 3.3 x 3.4
A368	226	8.1 x 3.3 x 3.4
B317	226	8.1 x 3.3 x 3.4
A	NC-Off	4.1 x 2.7 x 2.7
B	NC-Off	4.4 x 4.0 x 2.7
C	RTP	6.4 x 3.8 x 2.6
D	RTP	8.6 x 12.1 x 3.6

* Space height is the last dimension listed

Table 1 Test site, date, and room dimensions

Each of the four NIST rooms was tested twice. The first test was conducted to determine the leakage of the room with all ventilation systems off and all vents sealed. The second test was

12

conducted after sealing all visible cracks, outlets and penetrations. The four rooms located in North Carolina were all involved in a series of SIP tracer gas studies conducted by Jetter and Profitt [13] and were pressure tested only once as part of the current effort, under the same sealing conditions as in the referenced study. Note that Room B has one exterior wall with windows.

3.3 CONTAM Predictions of Protection Factor

In order to investigate the relationship between SIP zone leakage and Protection Factor, a series of simulations was performed using the multizone airflow and contaminant dispersal program CONTAM [34]. Three buildings were considered in order to develop a more complete understanding of the impact of various parameters on agent exposure within the SIP space: a simple one-story building, a simple ten-story building, and a more realistic two-story office building.

The simple one and ten story buildings were configured as a single open zone with the SIP space contained within that volume. The one-story building model has a floor area of 110 m^2 and a ceiling height of 3 m, whereas the ten-story model has a total floor area of 1010 m^2 and a building height of 30 m (3 m per floor). In both models the shelter floor area is 10 m^2 with a ceiling height of 3 m. These two simple models provide a first order sense of the impacts of shelter and building tightness.

The two-story office building model, depicted in Figure 9 as a CONTAM floor plan, includes restrooms, elevators, stairs, a lobby and a conference room. (The lines in the figure depict pressure differences and airflow rates from a set of CONTAM predictions.) The conference room, located on the second level, is designated as the SIP zone. Each level has a floor area of 920 m^2 and a ceiling height of 3 m. The shelter floor area is 260 m^2.

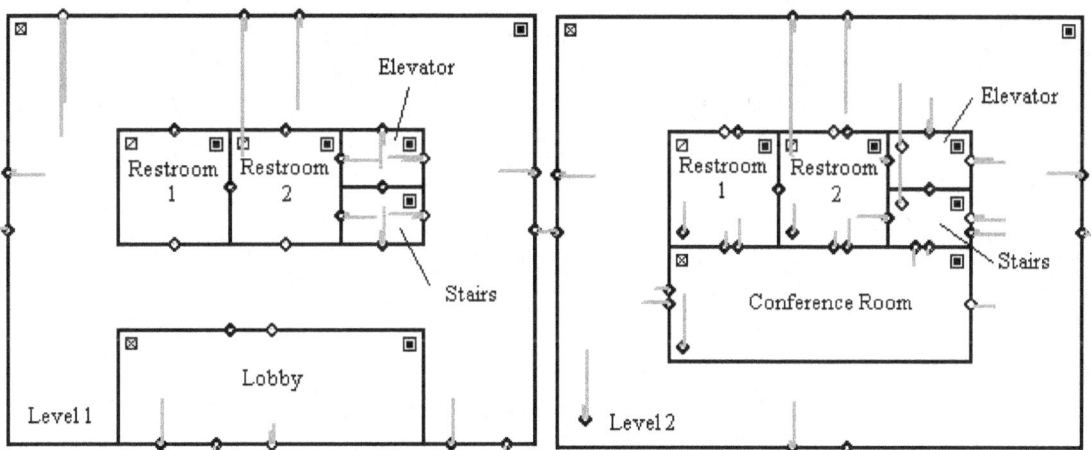

Figure 9 Two story building floor plan

The simulations are performed for two sets of weather conditions, cold and windy as well as mild and calm. The former conditions correspond to an indoor-outdoor temperature difference of 20 °C and a wind speed of 5 m/s, while the latter is defined by a temperature difference of 2.5 °C and a wind speed of 1 m/s. These two sets of conditions result in high and low building air change rates, which bound the results in terms of the amount of outdoor contaminant that enters the building and therefore is available to expose the shelter occupants. The building and shelter leakage values are also varied, including exterior wall ELA values from 1 cm^2/m^2 to 20 cm^2/m^2 and shelter wall ELA values from 0.1 cm^2/m^2 to 10 cm^2/m^2. The exterior wall leakage values are based on measurements conducted in a range of commercial buildings [40]. The shelter values are based on the limited data that exist for interior wall leakage [35] and the assumption that shelter walls will generally be tighter than exterior walls. Two values of shelter occupant density are employed in the simulations, 1 m^2 of floor area per person based on a minimum recommendation in several documents [2, 14], and 2 m^2 per person. The former value is consistent with FEMA recommendations for tornado shelters, where air leakage and contaminant entry are not issues. Based on potential concerns about CO_2 build up, the lower occupancy density is also considered, which happens to be ten times the default occupant density for office space in ASHRAE Standard 62.1-2007 [41].

The simulations employed an elevated outdoor concentration of 1.16 mg/m^3 (1 ppm(v)) lasting 1 min beginning one hour into the 2-day simulation period. The concentration in the shelter C_S is converted to occupant exposure E_S (in units of mg$\cdot$min/m^3) by integrating through some time t, as shown in Equation (7).

$$E_S = \int_0^t C_S dt \qquad (7)$$

Similarly the outdoor exposure, E_{OUT}, is based on the outdoor concentration C_{OUT} and is calculated using Equation (8),

$$E_{OUT} = \int_0^t C_{OUT} dt \qquad (8)$$

In the case of a constant outdoor concentration, the outdoor exposure is simply the concentration multiplied by the time over which it is elevated, i.e., 1.16 mg$\cdot$min/m^3.

As described above, the dimensionless Protection Factor PF_{OUT} is the outdoor exposure E_{OUT} divided by the shelter exposure E_S, which can be expressed as follows:

$$PF_{OUT} = \frac{E_{OUT}}{E_S} \qquad (9)$$

Alternatively, the Protection Factor can be based on the exposure in the rest of the building E_B defined as in Equation (7) but using the building concentration C_B instead of the shelter concentration. In that case, the Protection Factor is expressed as follows:

$$PF_B = \frac{E_B}{E_S} \qquad (10)$$

The CONTAM simulations were also used to predict CO_2 levels in the shelter, based on an assumed CO_2 generation rate of 0.0052 L/s for each person and occupant densities corresponding to 1 m^2/person and 2 m^2/person.

3.4 CONTAM Simulations of SIP Leakage Measurements

In considering measured shelter tightness as an evaluation criterion, the question arises of how well the measured value corresponds to the actual tightness. This question exists because a fan pressurization test of a single zone in a multizone building will not be able to achieve identical pressure differences across all partitions between the shelter and adjacent building zones. The level of uncertainty that this effect will cause is not clear, but needs to be examined. In this phase of the study, CONTAM is used to simulate fan pressurization tests of shelter airtightness in a multizone building. The airtightness value from the simulated tests is then compared to the airtightness value that is input into the CONTAM model of the building. This analysis enables the determination of uncertainty estimates for the field measurements.

3.5 Validation of Predictions from Shelter Airtightness Values

The use of shelter airtightness as a surrogate for Protection Factor relies on the existence of a reliable relationship between this airtightness value and the airflows between the outdoors, the building, and the shelter. It is beyond the scope of this study to conduct the experiments needed to perform a comprehensive validation by comparing measured and predicted airflows. However, as noted earlier, shelter pressure tests were conducted in four commercial building shelter spaces that were employed in an SIP tracer gas study conducted by EPA [13]. The results from that study, specifically shelter air change rate estimates, are compared with model predictions as a limited validation exercise to support the results of the current study.

4. STUDY RESULTS

This section presents the results of the measurements and analyses described in the previous section, including the fan pressurization tests of shelter airtightness, the CONTAM predictions of protection factor, the CONTAM simulations to assess potential measurement errors in the field, and validation of the predictions through comparison with limited field measurements.

4.1 SIP Airtightness

The results of the zone airtightness tests are presented in Table 2 for each of the eight rooms tested. As noted earlier, the NIST spaces were tested once with only limited sealing and again with extra sealing. Also, ELA values for the NIST spaces are presented for both zone pressurization and depressurization conditions. The four spaces in North Carolina were only tested under depressurization and only with sealing consistent with the test conditions used in the earlier study by Jetter and Profitt [13]. Under limited sealing the ELA values are in a relatively narrow range from somewhat above 1 cm^2/m^2 to about 5 cm^2/m^2. The sealed values are lower as expected and range from 0.25 cm^2/m^2 to just under 1 cm^2/m^2. In general, the pressurization and depressurization test values are similar, with the exception of Room 226/A368 with extra sealing in place. The percent reduction due to sealing is surprisingly consistent for the four spaces, ranging from about 60 % to 90 %.

The recent FEMA report on shelters and safe rooms cited earlier includes fan pressurization test results of a stairwell that was being considered as an SIP space and provides another airtightness data point [2]. The stairwell was approximately 43 m high and had a cross-section of 10.7 m by 3.7 m. The airtightness test was done once as is and again with the doors sealed. The as is and sealed ELAs at 4 Pa were 3.1 cm^2/m^2 and 2.1 cm^2/m^2, respectively. These values are in the range of the measurements presented in Table 2. This same FEMA report contains a table (3-2) with airflow values, in units of cfm per ft^2 of floor area, required to pressurize a room to 25 Pa. These values are presented for four levels of tightness, very tight, tight, typical and loose. Converting these values into the units presented in Table 2, assuming a room size of 5 m x 5 m x 2.5 m, the results are as follows: 0.47 cm^2/m^2, 2.36 cm^2/m^2, 5.89 cm^2/m^2 and 11.78 cm^2/m^2. These values of the FEMA report are certainly consistent with those measured in the eight spaces considered in this study.

| | Effective Leakage Area, ELA (cm^2/m^2) | | | | Percent change in ELA due to sealing | |
| | Limited Sealing | | Extra Sealing | | | |
Building/Room	Press	Depress	Press	Depress	Press	Depress
226/B221	4.80	5.08	0.94	0.98	80	81
226/B113	2.60	2.72	0.37	0.47	86	83
226/A368	1.74	1.86	0.71	0.37	59	80
226/A317	1.76	1.85	0.52	0.61	70	67
NC-Off/A	-	6.08	-	-	-	-
NC-Off/B	-	4.67	-	-	-	-
RTP/C	-	2.78	-	-	-	-
RTP/D	-	3.74	-	-	-	-
Mean	2.73	3.60	0.64	0.61	74	78
StdDev	1.44	1.56	0.25	0.27	12	7

Press and Depress indicates whether the SIP zone was under positive or negative pressure during the test.

Table 2: SIP airtightness test results

4.2 CONTAM Predictions of Protection Factor

This section presents the results of the CONTAM simulations that were conducted to investigate the relationship between shelter airtightness and protection factor. These simulations were performed in three model buildings, and predicted the concentrations of an agent released outdoors and CO_2 levels in the shelter due to the shelter occupants.

Figure 10 shows an example of the simulation results for the one-zone building model with a building leakage value of 5 cm^2/m^2 and a shelter leakage of 1 cm^2/m^2. The plot shows the agent concentration in the shelter over the 2 day (2880 min) simulation period, which starts to increase from 0 mg/m^3 starting at t = 1 h. The plot also contains the CO_2 concentration in the shelter, which increases from an initial value of 1800 mg/m^3. Note that the indoor CO_2 increases to fairly high levels, relative to the ACGIH short-term (1 min) exposure limit of 54 000 mg/m^3 and the threshold limit value (based on an 8 h exposures over a 40 h work week) of 9000 mg/m^3, which strongly suggests limits on the amount of time that such a shelter should be occupied [42].

16

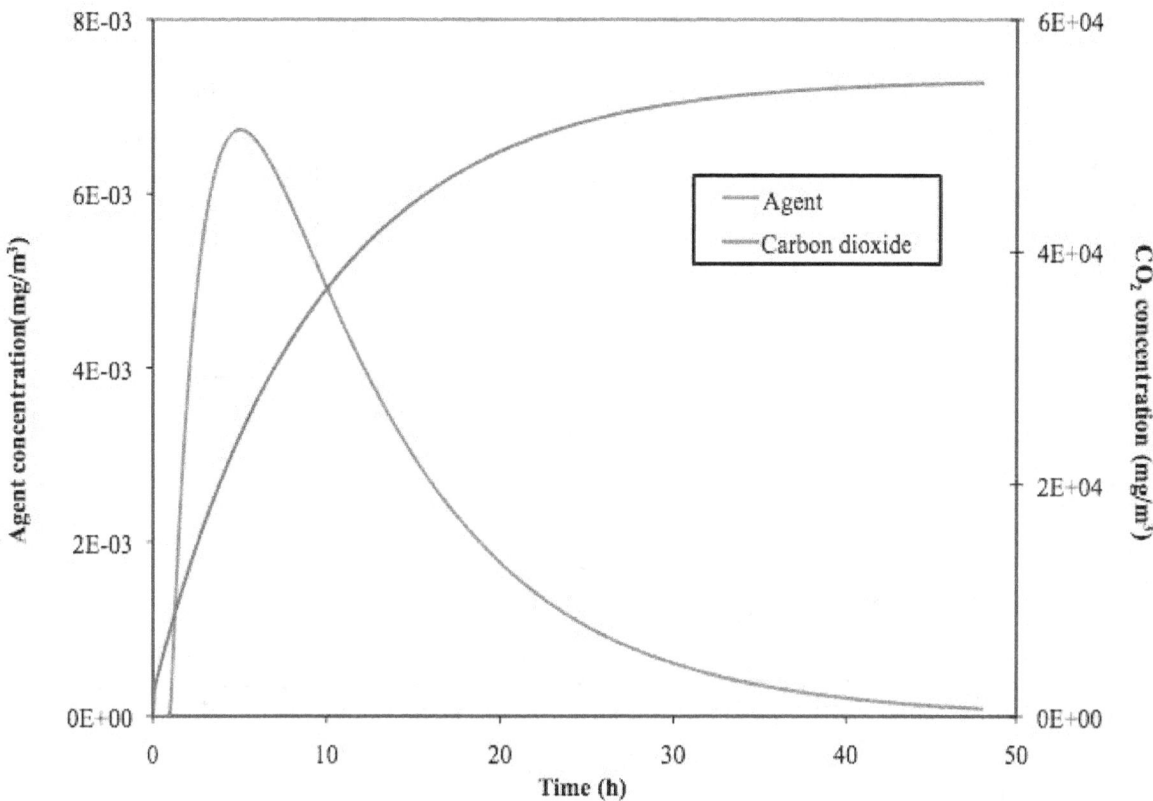

Figure 10 Example plot of simulated agent and CO_2 concentrations for the one-zone building

The simulations for other building and shelter leakage values and for the other buildings yield results similar to those seen in Figure 10. Tighter buildings and shelters tend to lower the peak agent concentration in the shelter, but also extend the period of time over which the agent remains in the shelter. For example, Figure 11 shows the agent concentrations for a building leakage value of 5 cm^2/m^2 and several values of the shelter leakage. As the shelter leakage increases, the peak concentration also increases but the duration of the elevated shelter concentration decreases. Increased tightness also increases the shelter CO_2 concentrations due to the decreased dilution rates.

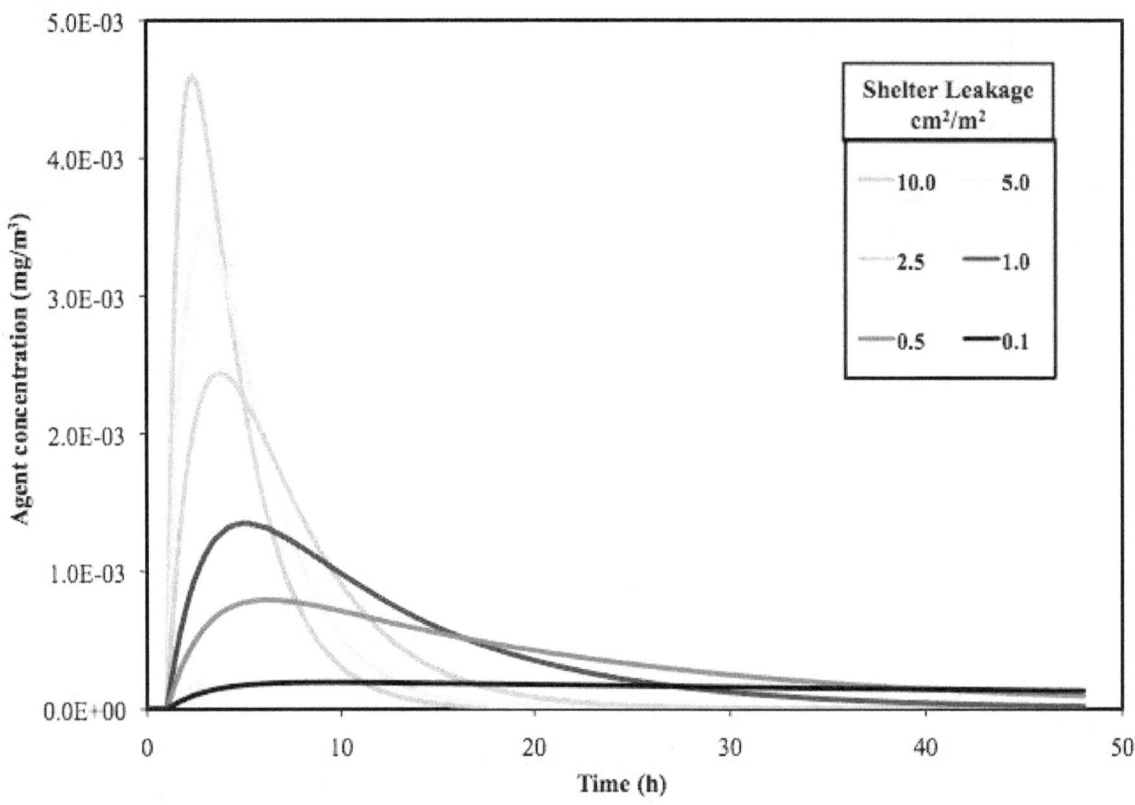

Figure 11 Simulated agent concentrations in shelter for multiple shelter leakage values

One-zone building model

The simulation results for the one-zone building model are shown in Tables 3 through 6. Table 3 shows protection factors after 1 h, 2 h and 3h for different combinations of building and shelter leakage under the cold and windy weather conditions, as well as the building and shelter air change rates for each combination of leakage values. The building air change rates depend only on the building leakage and range from 0.09 h^{-1} for the tightest building to 1.71 h^{-1} for the leakiest. The shelter air change rates describe the air change rate with the rest of the building and depend only on the shelter airtightness, covering a range of about 100 to 1 from the tightest to the leakiest shelter values.

Two different protection factors are presented in the table: PF_O, which is defined as the outdoor exposure divided by the exposure in the shelter after the designated time interval, and PF_B, which is the exposure in the rest of the building divided by the shelter exposure. The protection factors defined with reference to the building are always lower than the PF_O because the building concentration remains elevated after the outdoor concentration returns to zero. However, PF_B may be viewed as a better measure of protection for situations where the building occupants remain in the building during the sheltering period rather than go outdoors. The protection factors decrease over time, since the agent remains in the shelter after the outdoor episode is over. As noted earlier, the protection factor decreases as the building and shelter leakage increase. For a given duration of sheltering, the protection factors vary by more than two orders of magnitude, therefore, a very tight building and a very tight shelter can decrease the exposure in the shelter to less than 1 % of what it would be under the leakiest conditions.

18

Building leakage (cm²/m²)	Shelter leakage (cm²/m²)	Air change rate (h⁻¹)		Duration of sheltering (h)					
				1		2		3	
		Building	Shelter	PF_B	PF_O	PF_B	PF_O	PF_B	PF_O
1.0	0.1	0.09	0.01	182	2041	89	524	59	240
1.0	0.5	0.09	0.05	37	415	18	108	12	50
1.0	1.0	0.09	0.11	19	212	10	56	7	27
1.0	2.5	0.09	0.27	8	90	4	25	3	13
1.0	5.0	0.09	0.54	4	49	3	15	2	8
1.0	10.0	0.09	1.08	3	29	2	10	1	6
5.0	0.1	0.43	0.01	171	460	80	132	50	67
5.0	0.5	0.43	0.05	35	93	16	27	11	14
5.0	1.0	0.43	0.11	18	48	9	14	6	7
5.0	2.5	0.43	0.27	7	20	4	6	3	4
5.0	5.0	0.43	0.54	4	11	2	4	2	2
5.0	10.0	0.43	1.08	2	7	2	3	1	2
10.0	0.1	0.85	0.01	160	264	71	84	44	46
10.0	0.5	0.85	0.05	33	54	14	17	9	10
10.0	1.0	0.85	0.11	17	27	8	9	5	5
10.0	2.5	0.85	0.27	7	12	3	4	2	3
10.0	5.0	0.85	0.54	4	6	2	2	2	2
10.0	10.0	0.85	1.08	2	4	1	2	1	1
20.0	0.1	1.71	0.01	143	169	61	62	37	38
20.0	0.5	1.71	0.05	29	34	13	13	8	8
20.0	1.0	1.71	0.11	15	18	7	7	4	4
20.0	2.5	1.71	0.27	6	7	3	3	2	2
20.0	5.0	1.71	0.54	3	4	2	2	1	1
20.0	10.0	1.71	1.08	2	3	1	1	1	1

PF_B and PF_O are the protection factors with reference to the building and outdoor concentration, respectively.

Table 3: Protection factors for one-zone model (cold and windy weather)

Table 4 presents the air change rates and protection factors for the one-zone model for the mild and calm weather. The more temperate weather conditions reduce the building air change rates to about 25 % of their values under the cold and windy conditions, but do not impact the shelter-to-building air change rates. As a result, the values of PF_B do not change very much relative to the values in Table 3. However, PF_O increases by roughly a factor of four, corresponding approximately to the reduction in the building air change rate.

Building leakage (cm²/m²)	Shelter leakage (cm²/m²)	Air change rate (h⁻¹)		Duration of sheltering (h)					
		Building	Shelter	1		2		3	
				PF_B	PF_O	PF_B	PF_O	PF_B	PF_O
1.0	0.1	0.02	0.01	184	8516	92	2135	61	957
1.0	0.5	0.02	0.05	37	1731	19	441	13	201
1.0	1.0	0.02	0.11	19	883	10	229	7	106
1.0	2.5	0.02	0.27	8	374	4	103	3	50
1.0	5.0	0.02	0.54	4	205	3	61	2	32
1.0	10.0	0.02	1.08	3	122	2	41	1	23
5.0	0.1	0.10	0.01	181	1753	89	452	58	208
5.0	0.5	0.10	0.05	37	356	18	93	12	44
5.0	1.0	0.10	0.11	19	182	9	49	6	23
5.0	2.5	0.10	0.27	8	77	4	22	3	11
5.0	5.0	0.10	0.54	4	42	2	13	2	7
5.0	10.0	0.10	1.08	3	25	2	9	1	5
10.0	0.1	0.20	0.01	178	908	86	242	56	115
10.0	0.5	0.20	0.05	36	185	18	50	12	24
10.0	1.0	0.20	0.11	18	94	9	26	6	13
10.0	2.5	0.20	0.27	8	40	4	12	3	6
10.0	5.0	0.20	0.54	4	22	2	7	2	4
10.0	10.0	0.20	1.08	2	13	2	5	1	3
20.0	0.1	0.40	0.01	172	487	81	138	51	69
20.0	0.5	0.40	0.05	35	99	17	29	11	15
20.0	1.0	0.40	0.11	18	50	9	15	6	8
20.0	2.5	0.40	0.27	7	21	4	7	3	4
20.0	5.0	0.40	0.54	4	12	2	4	2	2
20.0	10.0	0.40	1.08	2	7	2	3	1	2

PF_B and PF_O are the protection factors with reference to the building and outdoor concentration, respectively.

Table 4: Protection factors for one-zone building model (mild and calm weather)

Figures 12 and 13 are graphical presentations of the data in Tables 3 and 4, with Figure 12 presenting the protection factor using the building exposure as a reference and Figure 13 based on the outdoor exposure. Such figures are used to present the results for the other building models, rather than using the format in Tables 3 and 4.

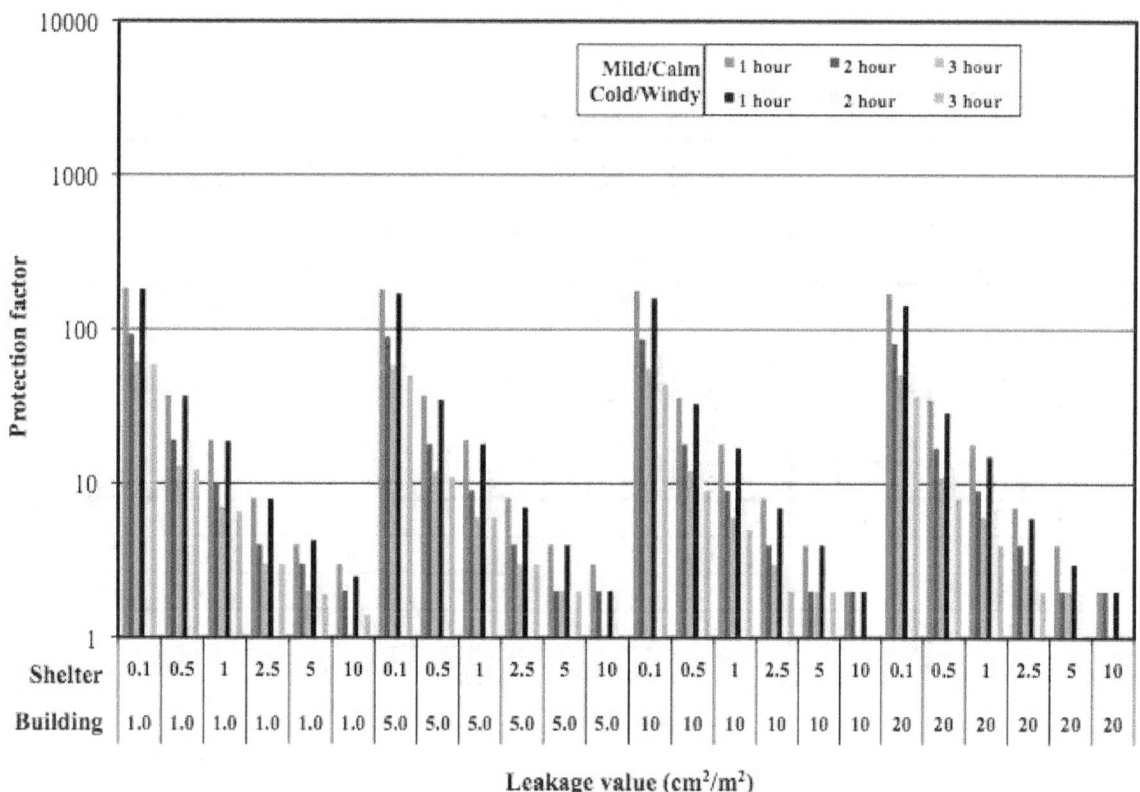

Figure 12 Protection factors, referenced to building, for one-zone model

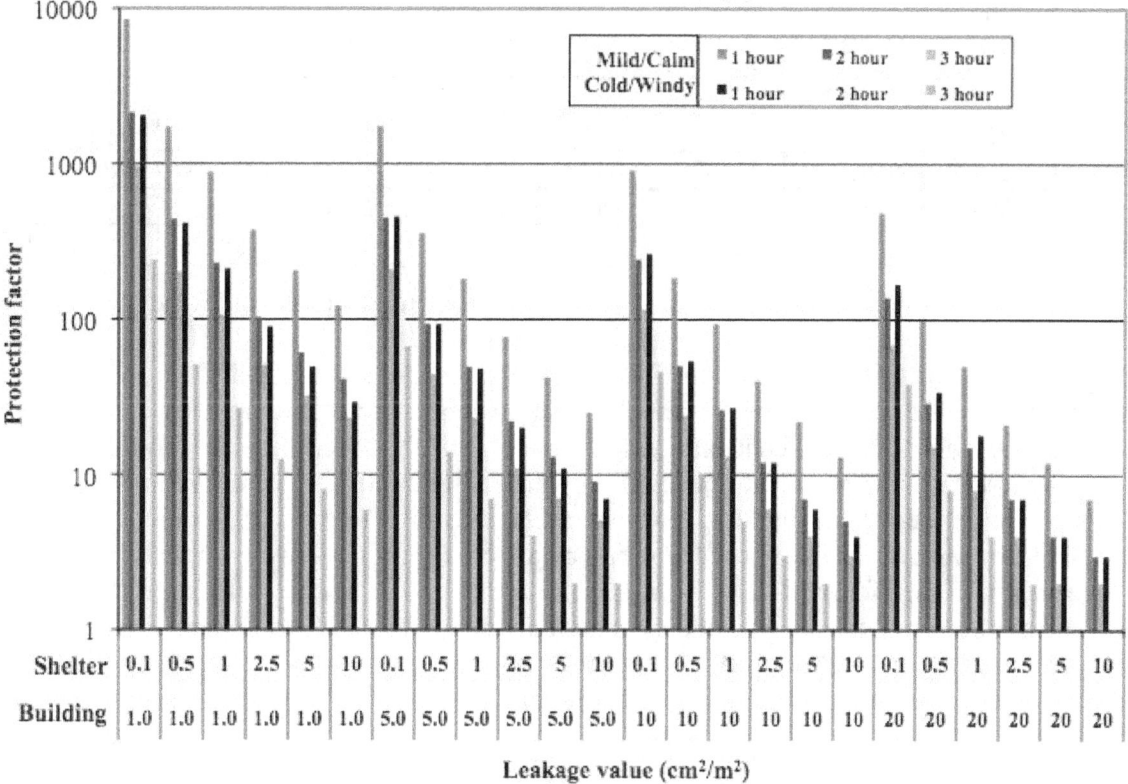

Figure 13 Protection factors, referenced to outdoors, for one-zone model

21

Building leakage (cm^2/m^2)	Shelter leakage (cm^2/m^2)	1 m^2 per occupant			2 m^2 per occupant		
		Duration of sheltering (h)			Duration of sheltering (h)		
		1	2	3	1	2	3
		Carbon dioxide concentration (mg/m^3)					
1.0	0.1	13364	24787	36085	7597	13308	18955
1.0	0.5	13113	23808	33933	7470	12814	17871
1.0	1.0	12811	22668	31514	7318	12239	16652
1.0	2.5	11973	19736	25721	6895	10761	13732
1.0	5.0	10776	16107	19430	6292	8932	10562
1.0	10.0	8965	11751	13184	5379	6736	7415
5.0	0.1	13362	24781	36073	7595	13301	18943
5.0	0.5	13104	23777	33874	7461	12783	17813
5.0	1.0	12793	22607	31397	7300	12180	16540
5.0	2.5	11929	19593	25443	6852	10625	13482
5.0	5.0	10695	15849	18929	6213	8696	10141
5.0	10.0	8824	11324	12377	5244	6367	6788
10.0	0.1	13360	24776	36066	7593	13297	18936
10.0	0.5	13095	23754	33837	7452	12760	17778
10.0	1.0	12775	22562	31325	7283	12136	16472
10.0	2.5	11888	19486	25271	6811	10525	13332
10.0	5.0	10618	15655	18618	6138	8522	9891
10.0	10.0	8690	11005	11878	5116	6098	6425
20.0	0.1	13358	24771	36060	7591	13292	18930
20.0	0.5	13084	23731	33809	7441	12738	17751
20.0	1.0	12752	22518	31268	7260	12093	16420
20.0	2.5	11833	19382	25132	6758	10428	13217
20.0	5.0	10518	15467	18363	6040	8358	9702
20.0	10.0	8517	10696	11475	4953	5853	6157

Table 5: Carbon dioxide concentrations for one-zone model (cold and windy weather)

Table 5 shows the CO_2 concentrations for the one-zone model at 1 h, 2 h and 3 h for two values of occupant density, 1 m^2 per person and 2 m^2 per person, under cold and windy conditions. Each value in the table is the CO_2 concentration in the shelter at the specified leakage values and time after sheltering commences. The concentrations are most sensitive to the duration of sheltering and fairly insensitive to the building leakage values. The three-to-one variation in sheltering duration leads to an increase in concentration by a factor of roughly two to three, which is similar to the variation seen for the 100-to-1 range in shelter leakage. The shelter leakage has more of an effect than the building leakage, but the sheltering duration is still the dominant factor. None of the cases are as high as the ACGIH short-term (1 min) exposure limit of 54 000 mg/m^3, but several exceed the threshold limit value of 9000 mg/m^3 [42]. For the sake of discussion, a reference value of 10,000 mg/m^3 is useful, but note that this value is not a health-based criteria or a "safe" concentration limit. The higher value of occupant density, 1 m^2 per occupant, results in roughly a doubling of the CO_2 concentration for the same leakage and duration values, with almost all of the values being above 10,000 mg/ m^3. For the lower occupant density (2 m^2 per occupant), the concentration after 1 h of sheltering never attains this reference concentration, even for the lowest leakage values. The 2 h CO_2 concentrations exceed that reference value for all but the leakiest shelters, although having a leaky shelter is counter to the goal of sheltering.

Table 6 shows the CO_2 concentrations for the one-zone model under the mild and calm weather conditions. These values exhibit the same strong dependence on duration of sheltering, with less of an effect of shelter leakage and very little dependence on building leakage, which was seen for the more severe weather conditions. The concentrations themselves are only slightly higher than those seen in Table 5 for the same leakage and duration values.

Building leakage (cm^2/m^2)	Shelter leakage (cm^2/m^2)	1 m^2 per occupant			2 m^2 per occupant		
		Duration of sheltering (h)			Duration of sheltering (h)		
		1	2	3	1	2	3
		Carbon dioxide concentration (mg/m^3)					
1.0	0.1	13364	24788	36088	7597	13309	18959
1.0	0.5	13115	23815	33950	7472	12822	17887
1.0	1.0	12815	22683	31547	7322	12254	16684
1.0	2.5	11982	19772	25799	6905	10796	13804
1.0	5.0	10794	16172	19572	6309	8992	10684
1.0	10.0	8996	11860	13414	5409	6832	7599
5.0	0.1	13364	24786	36084	7597	13307	18954
5.0	0.5	13113	23806	33929	7470	12812	17867
5.0	1.0	12810	22665	31507	7317	12236	16645
5.0	2.5	11970	19728	25704	6893	10754	13717
5.0	5.0	10772	16093	19401	6288	8919	10537
5.0	10.0	8958	11729	13137	5372	6717	7378
10.0	0.1	13363	24784	36079	7596	13305	18950
10.0	0.5	13110	23796	33908	7467	12802	17847
10.0	1.0	12804	22644	31465	7311	12216	16605
10.0	2.5	11957	19680	25606	6880	10708	13628
10.0	5.0	10747	16006	19223	6263	8839	10386
10.0	10.0	8914	11585	12851	5329	6591	7152
20.0	0.1	13362	24781	36073	7595	13302	18944
20.0	0.5	13105	23779	33877	7462	12785	17816
20.0	1.0	12794	22611	31403	7301	12184	16546
20.0	2.5	11932	19602	25459	6855	10634	13496
20.0	5.0	10701	15865	18958	6218	8711	10165
20.0	10.0	8834	11351	12424	5253	6390	6823

Table 6: Carbon dioxide concentrations for one-zone model (mild and calm weather)

Figure 14 is a plot of the concentration data in Table 5 for the cold and windy weather conditions. This graphical format will be used for the other building model results. Since the concentrations are very similar for the mild and calm conditions, a second plot is not included.

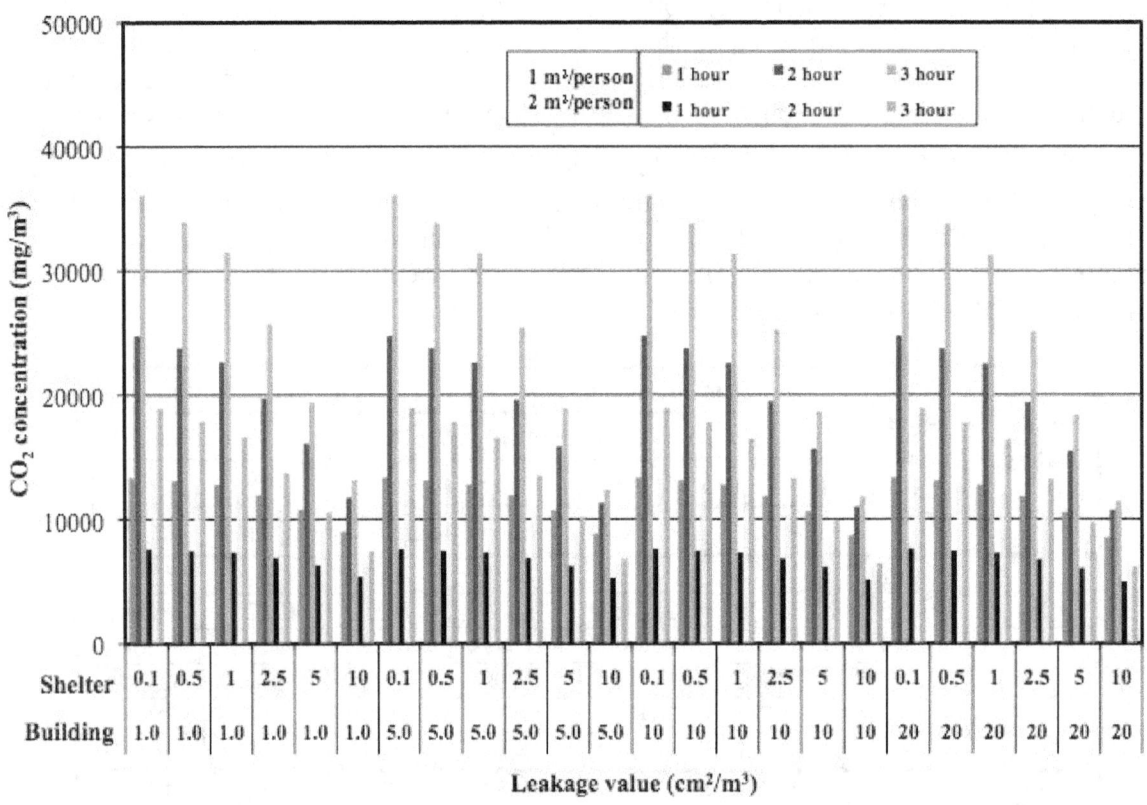

Figure 14 Shelter CO_2 concentration for one-zone model – cold/windy weather

Two-story office building model

Table 7 contains the air change rates for the two-story office building model as a function of building and shelter leakage and for the two sets of weather conditions. The building air change rates are similar in magnitude to those seen for the simple one-zone model. The shelter rates are consistently lower than for the one-zone model, as low as 20 % of the one-zone rates, but still on the same order of magnitude. The simulated protection factors are shown in Figures 15 and 16. (Note that these figures have a different scale than for the one-zone results in Figures 12 and 13.) Figure 15 shows the protection factor after 1 h, 2 h and 3 h for different combinations of building and shelter leakage using the building exposure as the reference. These building-based protection factors are almost an order of magnitude larger than those seen in the one-zone model for the mild weather conditions. This difference is due to the lower shelter air change rates. Under the cold-windy weather, the office building protection factors are somewhat higher than the one-zone model but the difference is smaller given that the air change rates are closer to those seen for the one-zone case. Figure 16 shows the protection factors referenced to the outdoor exposure. These protection factors are again well above those seen for the one-zone model due to the lower building and shelter air change rates. Overall these results display trends similar to those seen for the one-story building. Higher protection factors correspond to tighter buildings and tighter shelters, and shelter leakage has a more significant impact than building leakage.

24

Building leakage (cm^2/m^2)	Shelter leakage (cm^2/m^2)	Air change rate (h^{-1})			
		Cold and Windy		Mild and Calm	
		Building	Shelter	Building	Shelter
1.0	0.1	0.12	0.04	0.02	0.01
1.0	0.5	0.12	0.07	0.02	0.02
1.0	1.0	0.12	0.09	0.02	0.04
1.0	2.5	0.12	0.11	0.02	0.08
1.0	5.0	0.12	0.18	0.02	0.15
1.0	10.0	0.12	0.31	0.07	0.28
5.0	0.1	0.35	0.05	0.07	0.01
5.0	0.5	0.35	0.10	0.07	0.03
5.0	1.0	0.35	0.13	0.07	0.04
5.0	2.5	0.36	0.16	0.07	0.08
5.0	5.0	0.36	0.20	0.07	0.15
5.0	10.0	0.36	0.34	0.12	0.28
10.0	0.1	0.64	0.06	0.12	0.01
10.0	0.5	0.64	0.11	0.12	0.03
10.0	1.0	0.64	0.14	0.12	0.04
10.0	2.5	0.64	0.17	0.12	0.08
10.0	5.0	0.64	0.21	0.12	0.15
10.0	10.0	0.64	0.35	0.12	0.28
20.0	0.1	1.18	0.06	0.21	0.01
20.0	0.5	1.18	0.12	0.21	0.03
20.0	1.0	1.18	0.14	0.21	0.04
20.0	2.5	1.18	0.17	0.21	0.08
20.0	5.0	1.18	0.23	0.21	0.15
20.0	10.0	1.18	0.37	0.21	0.28

Table 7: Air change rates for two-story office building model

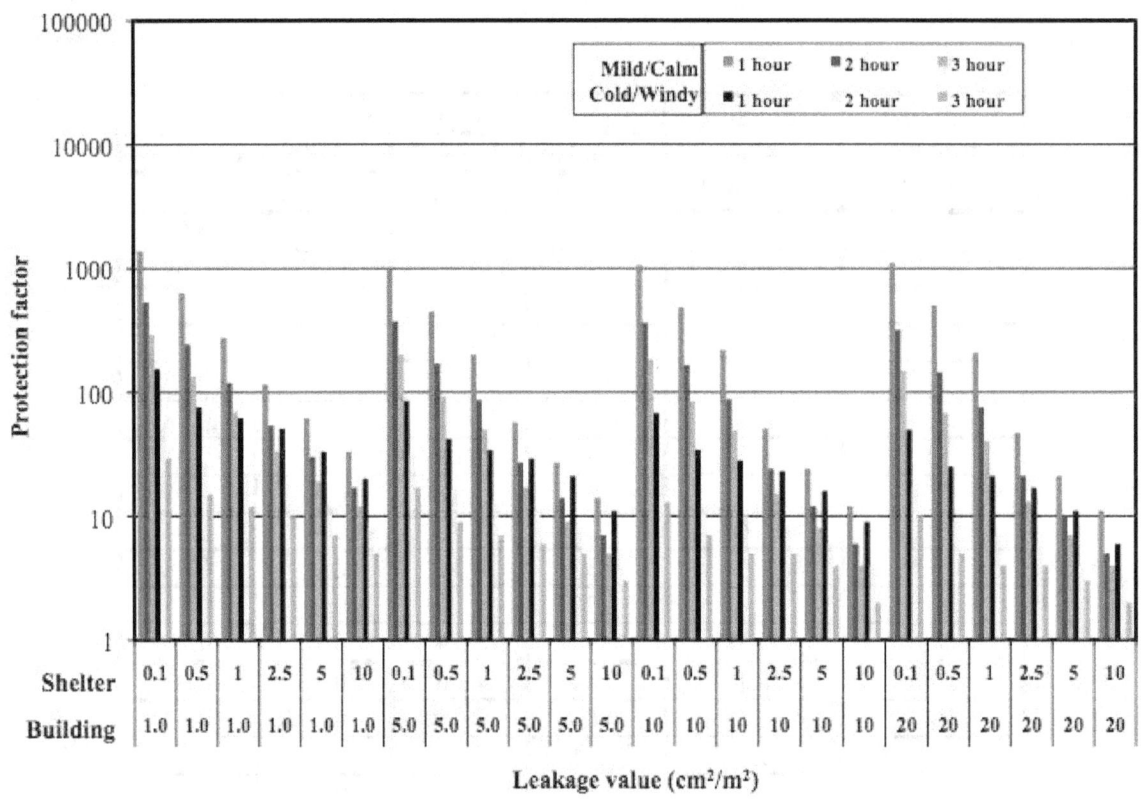

Figure 15 Protection factors, referenced to building, for office building model

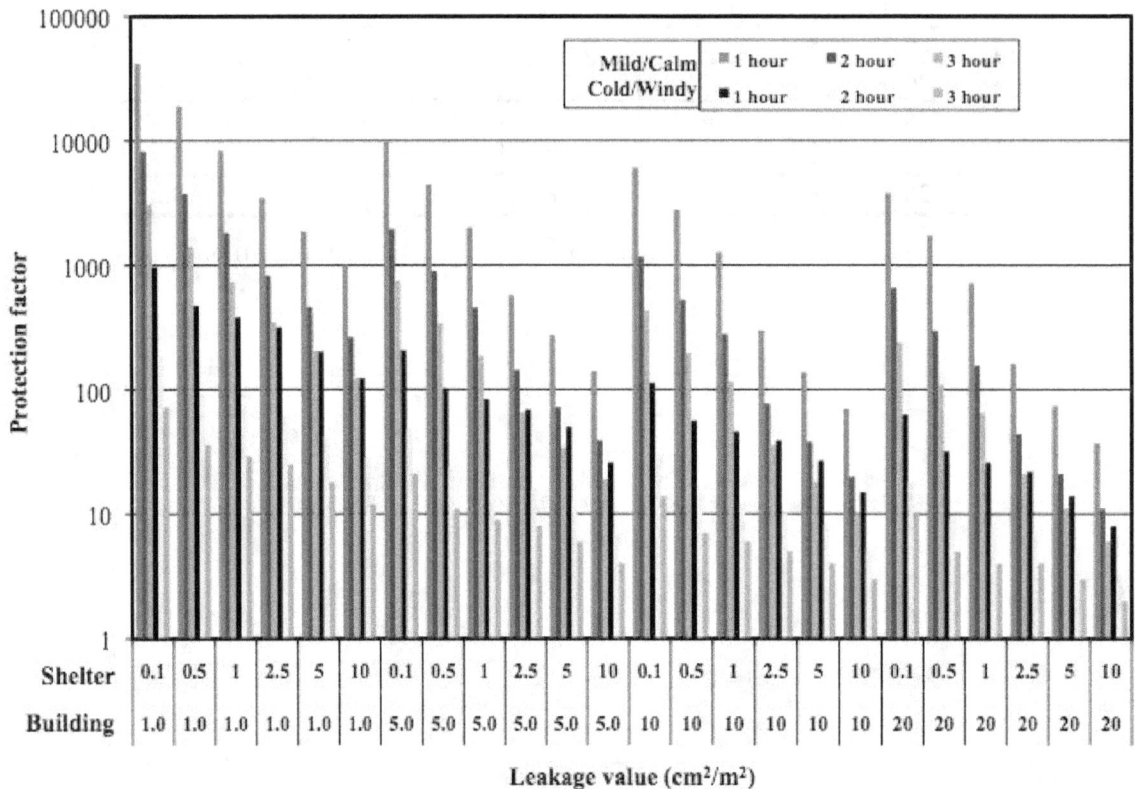

Figure 16 Protection factors, referenced to outdoors, for office building model

Figure 17 shows the CO_2 concentrations in the shelter at 1 h, 2 h and 3 h for the two-story office building. Concentrations are presented for the two different occupant densities and the various building and shelter leakage values. As in the case of the one-story model, the duration of sheltering is more significant than building or shelter tightness. The lower occupant density, 2 m^2 per person, reduces the concentrations significantly. The shelter concentrations are all above the reference value of 10 000 mg/m^3 except for the lower occupant density after 1 h of sheltering.

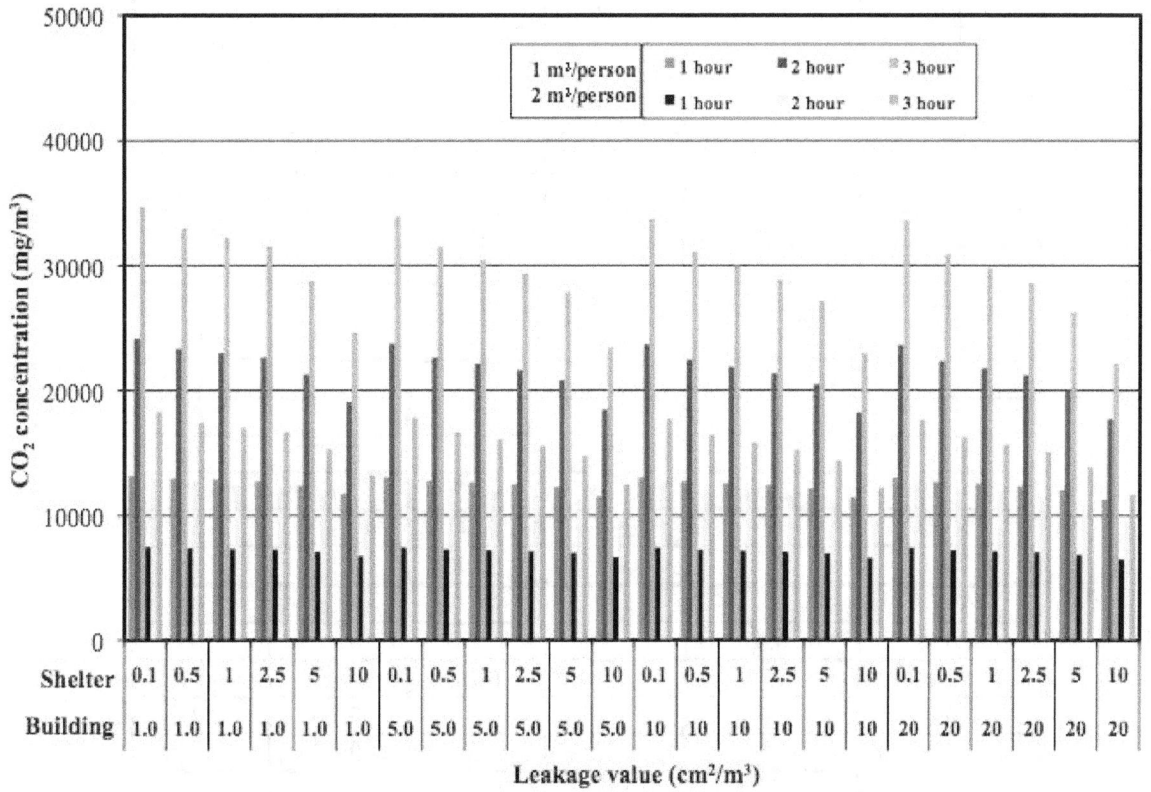

Figure 17 Shelter CO_2 concentration for office building model – cold/windy weather

Ten-story building model
Table 8 contains the air change rates for the ten-story building model as a function of building and shelter leakage and for the two sets of weather conditions. The values are very similar to those seen for the one zone model. The simulated protection factors are shown in Figures 18 and 19 for different combinations of building and shelter leakage, which are also very similar to those seen in the one-zone model. Given the configuration of the ten-story building, these results are not very surprising as the primary difference between this model and the one-zone is that the taller building has stronger stack (or temperature difference driven) pressures. However, the larger volume of the ten-story building leads to somewhat lower air change rates.

27

Building leakage (cm^2/m^2)	Shelter leakage (cm^2/m^2)	Air change rate (h^{-1})			
		Cold and Windy		Mild and Calm	
		Building	Shelter	Building	Shelter
1.0	0.1	0.08	0.01	0.02	0.01
1.0	0.5	0.08	0.05	0.02	0.05
1.0	1.0	0.08	0.11	0.02	0.11
1.0	2.5	0.08	0.27	0.02	0.27
1.0	5.0	0.08	0.54	0.02	0.54
1.0	10.0	0.12	1.08	0.07	1.08
5.0	0.1	0.38	0.01	0.09	0.01
5.0	0.5	0.38	0.05	0.09	0.05
5.0	1.0	0.38	0.11	0.09	0.11
5.0	2.5	0.38	0.27	0.09	0.27
5.0	5.0	0.38	0.54	0.09	0.54
5.0	10.0	0.38	1.08	0.09	1.08
10.0	0.1	0.76	0.01	0.17	0.01
10.0	0.5	0.76	0.05	0.17	0.05
10.0	1.0	0.76	0.11	0.17	0.11
10.0	2.5	0.76	0.27	0.17	0.27
10.0	5.0	0.76	0.54	0.17	0.54
10.0	10.0	0.76	1.08	0.17	1.08
20.0	0.1	1.52	0.01	0.35	0.01
20.0	0.5	1.52	0.05	0.35	0.05
20.0	1.0	1.52	0.11	0.35	0.11
20.0	2.5	1.52	0.27	0.35	0.27
20.0	5.0	1.52	0.54	0.35	0.54
20.0	10.0	1.52	1.08	0.35	1.08

Table 8: Air change rates for ten-story building model

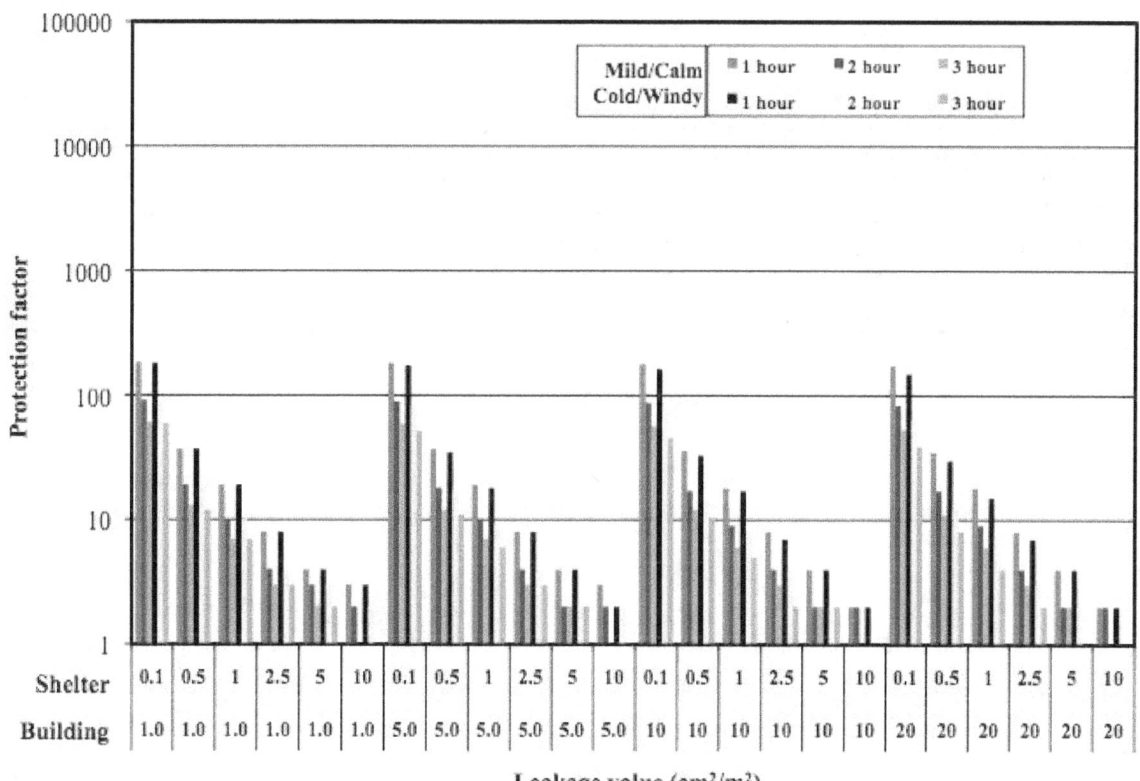

Figure 18 Protection factors, referenced to building, for ten-story building model

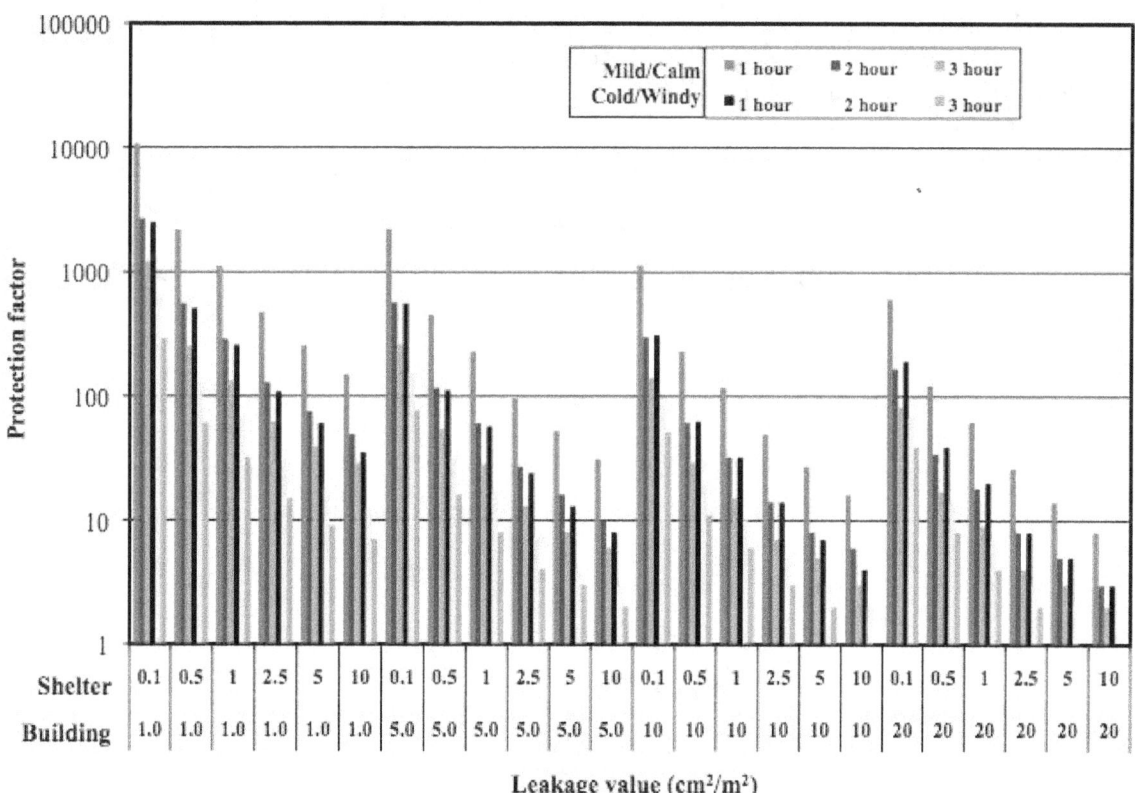

Figure 19 Protection factors, referenced to outdoors, for ten-story building model

Figure 20 shows the CO_2 concentrations in the shelter at 1 h, 2 h and 3 h for the ten-story office building. Concentrations are presented for the two different occupant densities and the various building and shelter leakage values. As with the other models, the duration of sheltering is more significant than building or shelter tightness and the lower occupant density reduces the concentrations significantly. The shelter concentrations are all above the reference value of 10 000 mg/m^3 except for the lower occupant density after only 1 h of sheltering.

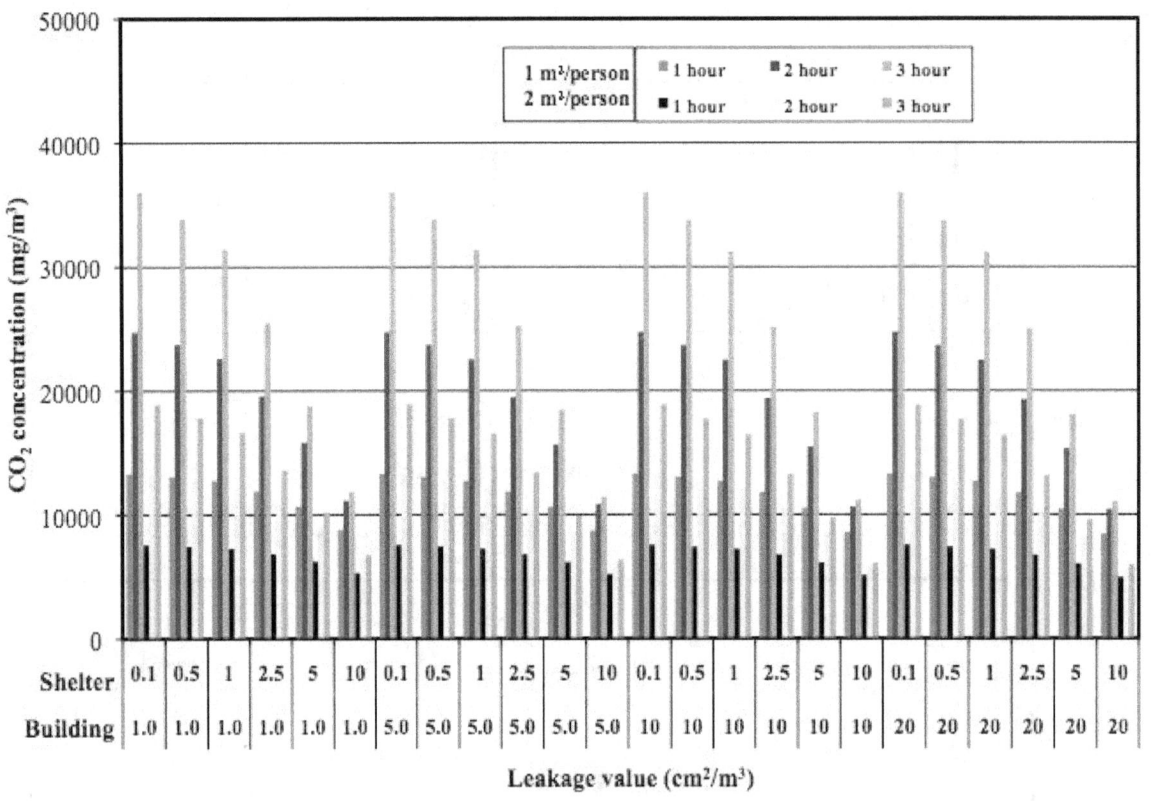

Figure 20 Shelter CO_2 concentration for ten-story model – cold/windy weather

4.3 CONTAM simulations of SIP leakage measurement

In order to investigate the impacts of interzone airflows and pressures on the ability to reliably estimate shelter leakage using a pressurization test, a series of CONTAM [34] simulations was performed for a ten-story building model. As described below, this ten-story model is different from the model used in the protection factor calculations described previously. In these simulations, a pressurization test to measure shelter airtightness was simulated with CONTAM by imposing an airflow on the SIP zone and then recording the airflow rates out of the zone and zone-to-hallway pressure differences as determined by the simulations. Figure 21 shows the CONTAM sketchpad for one floor of the building, with the SIP zone highlighted in green in the center of the floor, surrounded by five other zones. The colored zones at the upper edge of the floor plan represent stairway and elevator shafts.

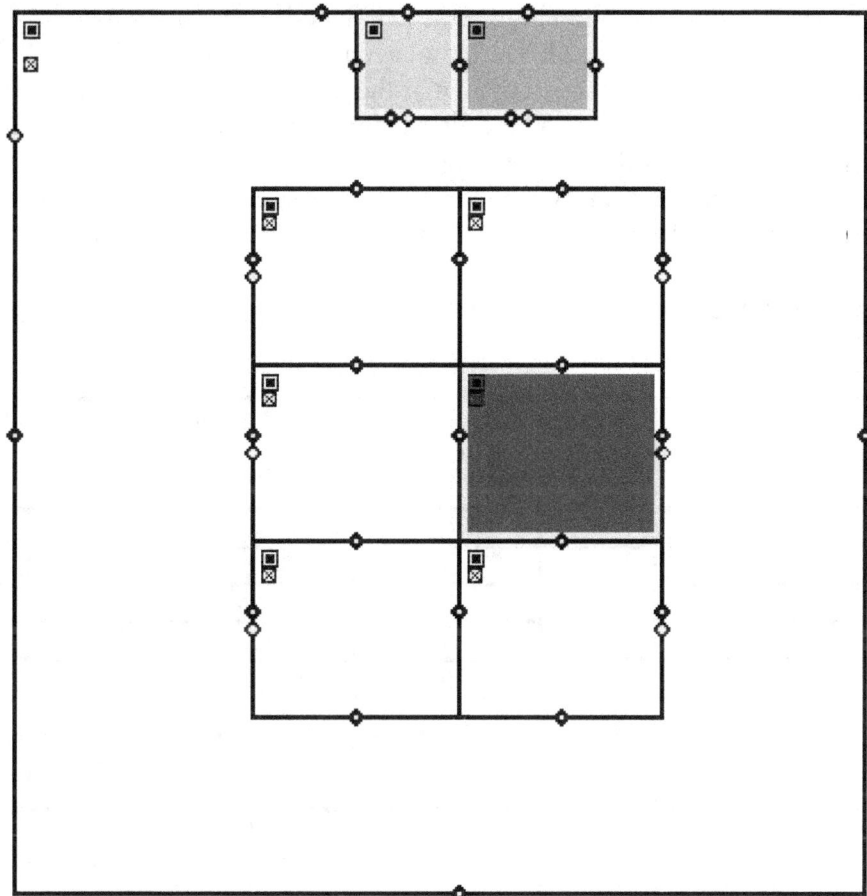

Figure 21 CONTAM sketchpad of simulated pressurization tests

The goal of these simulations was to determine if erroneous leakage values might result if the SIP zone being tested shared walls with other building zones, since the pressure difference measured during a test is typically from the SIP zone to the hallway. In addition, the impacts of ventilation system airflows in other zones were also examined, as it may not always be feasible to deactivate building ventilation systems during these tests. Note that system airflows to the SIP zone itself would need to be eliminated by turning off fans or sealing vents. Several such simulations were conducted for SIP zones on the first, fifth and ninth floor of the ten-story building. Simulations were conducted with the ventilation system off and with 10 % more supply airflow than return airflow to those building not being tested. Note that there are no system flows into the SIP zone itself during these simulations. In addition, pressurization tests were simulated with the SIP zone pressurized and depressurized relative to the building hallway. Finally, these simulations were conducted under four different weather conditions: zero wind speed and 0 °C indoor-outdoor temperature difference (ΔT); 5 m/s wind and 0 °C ΔT; 0 m/s wind speed and 20 °C ΔT; and 5 m/s wind speed and 20 °C ΔT.

Table 9 shows the simulation results in terms of the effective leakage area at 4 Pa calculated from the simulated pressure test. The actual leakage values input into the CONTAM model of the building are listed in the last row of the table. The simulated tests on the first floor with the system off are very close to the model value for all weather conditions. When the system is on, the simulated test results are about 10 % high if the SIP zone is being pressurized during the test and 10 % low if it is being depressurized. The simulated test results on the fifth and ninth floors are more sensitive to weather condition, in particular the outdoor air temperature. With a 20 °C indoor-outdoor temperature difference and the air handler off, the simulated test results are 15 % to 20 % low when the SIP zone is being pressurized by the test and a similar magnitude high when the zone is being depressurized. When the building air handler is on, the simulated value is about 10 % high when the SIP zone is pressurized and about 10 % low when it is depressurized. Therefore, interzone, system operation and weather effects are expected to lead to roughly 10 % errors in SIP pressurization test results, with worse case conditions leading to perhaps 20 % errors.

Simulation condition	Wind speed (m/s)	Temperature difference (°C)	Effective leakage area (cm^2)		
			First floor	Fifth floor	Ninth floor
Air handler off, pressurized	0	0	130.0	176.4	176.4
	0	20	130.2	151.4	146.7
	5	0	130.0	176.3	174.6
	5	20	130.2	150.6	143.5
Air handler on, pressurized	0	0	139.0	193.7	192.6
	0	20	140.5	174.8	166.5
	5	0	139.2	193.7	189.1
	5	20	140.6	176.9	163.5
Air handler off, depressurized	0	0	130.1	176.5	176.5
	0	20	129.9	194.6	196.7
	5	0	130.1	176.4	177.6
	5	20	129.9	195.1	197.9
Air handler on, depressurized	0	0	118.7	155.2	156.5
	0	20	117.2	173.8	178.1
	5	0	118.6	154.9	160.6
	5	20	117.1	171.7	179.6
Actual value in model		--	130.1	176.5	176.5

Table 9: Results of simulated SIP tests

4.4 Comparison of Predictions to Measurements

While it is beyond the scope of this study to conduct the experiments needed to perform a comprehensive validation of the CONTAM predictions presented in Section 4.2, a limited validation exercise is possible based on four of the tested shelter spaces being part of a tracer gas study conducted by EPA [13]. In particular, that study involved the measurement of tracer gas decay rates in the four shelters, which can be compared with the rates predicted from the CONTAM simulations. Table 10 shows the decay rates measured in the four spaces along with the air change rates predicted from the CONTAM simulations. The predicted rates are based on the measured shelter tightness values for each space, extrapolating between the airtightness values used in the predictions. The one-zone model predictions are used for the NC building spaces based on the layout of that building, while the 2-story model is used for the two spaces in the more complex RTP building. In the latter case, predictions are presented for both the cold/windy and calm/mild weather conditions, as the actual weather conditions are not reported with the measurements. While the comparison is very limited, the measured and predicted values are in reasonable agreement considering the many unknowns impacting the predictions and the uncertainties in the measured values.

Building/Room	Measured tracer gas decay rate from Ref [13] (h^{-1})	Predicted air change rate (h^{-1})
NC-Off/A	0.72	0.66
NC-Off/B	0.47	0.51
RTP/C 2.8	0.18	0.17, 0.09*
RTP/D 3.7	0.09	0.19, 0.11*

* Values predicted for cold/windy and mild/calm conditions.

Table 10: Comparison of measured and predicted decay rates

5. DISCUSSION

The purpose of this project was to develop and demonstrate methods to assess shelter airtightness in relation to the protection shelters provide in the event of outdoor contaminant releases. In actual application, a building owner or manager would select spaces for use as shelters based on a number of qualitative considerations identified previously and perhaps make modifications to increase the degree of protection offered by the shelter. In the case of unventilated shelters intended for short term sheltering, which are the subject of this study, a key modification is to increase the airtightness of the shelter through sealing of the boundaries to adjacent spaces. This project has focused on the relationship of shelter and building airtightness to the protection provided, with space pressurization testing examined as an evaluation method.

Note that this effort focused on unventilated shelters, in which building occupants are expected to reside for only an hour or two. When shelters are ventilated and employ air-cleaning and space pressurization strategies, occupants can stay in the shelter for longer periods of time because the level of CO_2 will not increase as much or as quickly, nor the level of oxygen decrease. Space airtightness is still important as it impacts the airflow required to pressurize the space, and pressurization testing is still the method of choice for quantifying space airtightness. Note also that there is a wealth of guidance on these more sophisticated, longer-term SIP approaches [2], and this study does not address ventilation requirements, air cleaning equipment and other issues related to ventilated shelters. Other important considerations such as structural design, communications, signage, medical care, decontamination of people or the building, and community planning are addressed in many of the documents cited in this report.

In reviewing existing SIP guidance, the lack of quantitative guidance on the shelter airtightness is notably lacking. There are many recommendations to tighten shelters, but no information on how tight or how to assess that tightness. As shown in this study, while tighter shelters (and tighter buildings) result in better protection against outdoor releases, they also limit the duration of occupancy in the shelter due to CO_2 buildup. Therefore, one must balance these benefits and concerns when considering shelter tightness and the use of unventilated shelters.

Room pressurization testing was seen to be relatively straightforward as applied in this study. While it is based on a standard test method for whole building airtightness testing, its application to individual rooms has yet to be standardized. Nevertheless, the simulations in this report showed that interzone pressure effects and system operation in non-test zones impacted the measurement results by about 10 %. The measured values of airtightness were surprisingly consistent among shelters tested, as well as the percentage increase in airtightness through sealing. Under limited sealing the ELA values were in a relatively narrow range from somewhat above 1 cm^2/m^2 to about 5 cm^2/m^2. The sealed values were lower as expected and ranged from 0.25 cm^2/m^2 to just under 1 cm^2/m^2. The percent reduction due to sealing was surprisingly consistent for the four spaces, ranging from about 60 % to 90 %.

The simulation of occupant exposure showed that tighter buildings and tighter shelters reduce exposure of SIP occupants, with shelter tightness having a greater impact than building tightness. Longer duration sheltering reduced the protection factor as expected, which highlights the importance of obtaining and communicating reliable information on when the outdoor hazard has ended and it is time to end the sheltering period. Based on these simulation results, CO_2 buildup over time may be more critical than the reduction in exposure. For the building and shelter airtightness values considered, the duration of sheltering was seen to be more important to the shelter CO_2 level than airtightness. None of the predicted CO_2 concentrations were as high as the ACGIH short-term (1 min) exposure limit of 54 000 mg/m^3, but several exceed the threshold limit value (TLV), based on 8 h exposure over 40 h work week, of 9000 mg/m^3 [42]. The TLV is not really relevant to a 1 h or 2 h exposure, but the high values seen in the simulations are of potential concern. Occupant density, or floor area per shelter occupant, is obviously an important determinant of these CO_2 concentrations. Many guides recommend 1 m^2 per occupant, but this value produced high CO_2 concentrations for many of the simulations in this study. A lower value of occupant density, e.g., 2 m^2 per occupant, might provide more habitable conditions for longer periods of time and merits consideration in future recommendations.

While the simulation results provided some useful insights, the results are highly dependent on the building models employed, including layout, airtightness, outdoor weather, and indoor temperatures. However, the relative results between the simulated cases are less sensitive to these factors than the absolute results.

This study does suggest some additional work to advance these results in the future. This additional work would include more field testing of different SIP spaces and the development of a standard test protocol. Ultimately, these efforts could support definitive airtightness criteria for unventilated shelters. The results of this study support a target airtightness value of 1 cm^2/m^2 and limiting unventilated sheltering to 2 h at the most, but these suggestions should not be considered as universally applicable recommendations.

The results of this study could be further supported by tracer gas studies. These studies could simulate an outdoor release and then be used to calculate actual protection factors for comparison to simulation results. In this way, the relationship between shelter airtightness and building protection could be more reliably demonstrated. In addition, it may be possible to develop tracer gas methods to determine Protection Factors directly without the intermediate step of calculating interzone airflows. These methods would involve releasing a tracer gas into a shelter space and then monitoring its decay over time, which in turn could potentially be related to the protection factor in the face of an outdoor release. While such a method has not been developed and would require a much higher level of expertise than pressurization testing, such an approach may be more accurate than an estimate of protection based on shelter airtightness.

6. ACKNOWLEDGEMENTS

This effort was supported by the U.S. Environmental Protection Agency under Interagency Agreement No. DW-13-92178301-0, but was not subjected to EPA peer review. The conclusions in this paper are therefore those of the authors and are not necessarily those of the U.S. EPA. The authors express their appreciation to John Chang, Jacky Rosati and Jim Jetter at EPA for their assistance in this project.

8. REFERENCES

[1] NIOSH. 2002. Guidance for Protecting Building Environments from Airborne Chemical, Biological, or Radiological Attacks. National Institute of Occupational Safety and Health, DHHS (NIOSH) Publication No. 2002-139.

[2] FEMA. 2006. Design Guidance for Shelters and Safe Rooms. Federal Emergency Management Agency Report No. FEMA 453.

[3] Janney, C., Janus, M., Saubier, L.F. and Widder, J. 2000. System Effectiveness Test of Home/Commercial Portable Room Air Cleaners. Battelle Report to U.S. Army Solder, Biological Chemical Command (SBCCOM), Contract No. SPO900-94-D-0002.

[4] Cristy, G.A. and Chester, C.V. 1981. Emergency Protection from Aerosols. Oak Ridge National Laboratory, Report No. ORNL-5519.

[5] Blewett, W.K. and Arca, V.J. 1999. Experiments in Sheltering in Place: How Filtering Affects Protection Against Sarin and Mustard Vapor. U.S. Army Edgewood Chemical Biological Center, Report No. ECBC-TR-034.

[6] Fradella, J. and Siegel, J.A. 2005. An Evaluation of Shelter-in-Place Strategies in Four Industrial Buildings. Proceedings of 10th International Conference on Indoor Air Quality and Climate: 3360-3364.

[7] Engelmann, R.J. 1990. Effectiveness of Sheltering in Buildings and Vehicles for Plutonium. U.S. Department of Energy, Report No. DOE/EH-0159T UC-160.

[8] Engelmann, R.J. 1992. Sheltering Effectiveness Against Plutonium Provided by Buildings, Atmospheric Environment, 26A (11): 2037-2044.

[9] Sohn, M.D., Sextro, R.G. and Lorenzetti, D.M. 2005. Assessing Sheltering-in-Place Responses to Outdoor Toxic Releases. Indoor Air, 15(11): 1792-1796.

[10] Rogers, G.O., Watson, A.P., Sorensen, J.H., Sharp, R.D. and Carnes, S.A. 1990. Evaluating Protective Actions for Chemical Agent Emergencies. Oak Ridge National Laboratory, Report No. ORNL-6615.

[11] Yuan, L.L. 2000. Sheltering Effects of Buildings from Biological Weapons, Science & Global Security, 8: 287-313.

[12] Jetter, J.J. and Whitfield, C. 2005. Effectiveness of Expedient Sheltering in Place in a Residence. Journal of Hazardous Materials, A119: 31-40.

[13] Jetter, J. and Proffitt, D. 2006. Effectiveness of Expedient Sheltering in Place in Commercial Buildings, Journal of Homeland Security and Emergency Management, 3 (2).

[14] Blewett, W.K., Reeves, D.W., Arca, V.J., Fatkin, D.P. and Cannon, B.D. 1996. Expedient Sheltering in Place: An Evaluation for the Chemical Stockpile Emergency Preparedness Program. Chemical Research, Development & Engineering Center, Report No. ERDEC-TR-336.

[15] Sorenson, J.H. and Vogt, B.M. 2001. Expedient Respiratory and Physical Protection: Does a Wet Towel Work to Prevent Chemical Warfare Agent Vapor Infiltration? Oak Ridge National Laboratory, Report No. ORNL/TM-2001/153.

[16] Sorenson, J.H. and Vogt, B.M. 2001. Will Duct Tape and Plastic Really Work? Issues Related to Expedient Shelter-In-Place. Oak Ridge National Laboratory, Report No. ORNL/TM-2001/154.

[17] CSEPP. 2001. Report of the Shelter-in-Place Work Group. Chemical Stockpile Emergency Preparedness Program.

[18] Shumpert, B. 2003. Questions and Answers Regarding Actions to Take When Ending Shelter-In-Place. Oak Ridge National Laboratory, Report No. ORNL/TM- 2003/230.

[19] Sorenson, J., Shumpert, B. and Vogt, B. 2002. Planning Protective Action Decision-Marking: Evacuate or Shelter-In-Place? Oak Ridge National Laboratory, Report No. ORNL/TM-2002/144.

[20] Yantosik, G.D., Lerner, K., Maloney, D. and Wasmer, F. 2001.When and How to End Shelter-In-Place Protection From a Release of Airborne Hazardous Material: Report on a Decision-Making Concept and Methodology. Argonne National Laboratory.

[21] Yantosik, G.D., Lerner, K. and Maloney, D.M., Temporary Shelter-In-Place as Protection Against a Release of Airborne Hazardous Material: Report of a Literature Search. 2001. Argonne National Laboratory.

[22] Yantosik, G.D., Maloney, D. and Wasmer, F., Comparison of Two Concepts and Methods to Decide When to End Temporary Shelter-in-Place Protection. 2003. Argonne National Laboratory.

[23] Vogt, B.M., Hardee, H.K., Sorensen, J.H. and Shumpert, B.L. 1999. Assessment of Housing Stock Age in the Vicinity of Chemical Stockpile Sites. Oak Ridge National Laboratory, Report No. ORNL/TM- 13742.

[24] FEMA. 1999. National Performance Criteria for Tornado Shelters. Federal Emergency Management Agency, Report No. 361.

[25] Chester, C.V. 1988. Technical Options for Protecting Civilians from Toxic Vapors and Gases. Oak Ridge National Laboratory, Report No. ORNL/TM- 10423.

[26] Mannan, M.S. and Kilpatrick, D.L., The Pros and Cons of Shelter-in-Place, Process Safety Progress. 19 (4) 210-218 (2000).

[27] NICS. 1999. Shelter in Place at Your Office. National Institute of Chemical Studies.

[28] NICS. 2001. Sheltering in Place as a Public Protective Action. National Institute of Chemical Studies.

[29] AFCES. 2001. Protective Actions for A Hazardous Material Release. Air Force Civil Engineer Support Agency, Tyndall Air Force Base.

[30] Chan, W.R. and Gadgil, A.J. 2004. Sheltering in Buildings from Large-Scale Outdoor Releases. Air Infiltration and Ventilation Centre, Report VIP No. 10.

[31] U.S. Army Corps of Engineers. 1998. Design of Chemical Agent Collective Protection Shelters for New and Existing Facilities. Report No. ETL 1110-3-490.

[32] U.S. Army Corps of Engineers. 1999. Design of Chemical Agent Collective Protection Shelters to Resist Chemical, Biological, and Radiological (CBR) Agents. Report No. ETL 1110-3-498.

[33] Price, P.N., Sohn, M.D., Gadgil, A.J., Delp, W.W., Lorenzetti, D.M., Finlayson, E.U., Thatcher, T.L., Sextro, R.G., Derby, E.A. and Jarvis, S.A. 2003. Protecting Buildings From a Biological or Chemical Attack: Actions to take before or during a release. Lawrence Berkeley National Laboratory, Report No. LBNL-51959.

[34] Walton, G.N. and Dols, W.S. 2005. CONTAMW 2.4 User Guide and Program Documentation. National Institute of Standards and Technology, Report No. NISTIR 7251.

[35] Persily, A.K. and Ivy, E.M. 2001. Input Data for Multizone Airflow and IAQ Analysis. National Institute of Standards and Technology, Report No. NISTIR 6585.

[36] ASTM. 2003. Standard E779-03, Standard Test Method for Determining Air Leakage Rate by Fan Pressurization. American Society for Testing and Materials.

[37] ASTM. 2007. Standard E1827-96 (2007), Standard Test Method for Determining Airtightness of Buildings Using an Orifice Blower Door. American Society for Testing and Materials.

[38] ASTM. 2008. Standard E1258-88 (2008), Standard Test Method for Airflow Calibration of Fan Pressurization Devices. American Society for Testing and Materials.

[39] ASHRAE. 2005. Fundamentals Handbook. American Society of Heating, Refrigerating and Air-Conditioning Engineers.

[40] Emmerich, S.J. and Persily, A.K. 2005. Airtightness of Commercial Buildings in the U.S. Proceedings of 26th AIVC Conference: 65-70.

[41] ASHRAE. 2007. ANSI/ASHRAE Standard 62.1-2007, Ventilation for Acceptable Indoor Air Quality. American Society of Heating, Refrigerating and Air-Conditioning Engineers.

[42] ACGIH. 2001. 2001 Threshold Limit Values (TLVs) for Chemical Substances and Physical Agents and Biological Exposure Indices (BEIs). American Conference of Governmental Industrial Hygienists.